Sophia Studies in Cross-cultural Philosophy of Traditions and Cultures

Volume 12

The Sophia Studies in Cross-cultural Philosophy of Traditions and Cultures fosters critical and constructive engagement of the intellectual and philosophical dimensions—broadly construed—of religious and cultural traditions around the globe. The series invites innovative scholarship, including feminist, postmodern, and postcolonial approaches.

More information about this series at http://www.springer.com/series/8880

Meera Baindur

Nature in Indian Philosophy and Cultural Traditions

Springer

Meera Baindur
Manipal Centre for Philosophy and Humanities
Manipal University
Manipal, Karnataka
India

ISSN 2211-1107 ISSN 2211-1115 (electronic)
Sophia Studies in Cross-cultural Philosophy of Traditions and Cultures
ISBN 978-81-322-2357-3 ISBN 978-81-322-2358-0 (eBook)
DOI 10.1007/978-81-322-2358-0

Library of Congress Control Number: 2015934684

Springer New Delhi Heidelberg New York Dordrecht London

Printed on acid-free paper

Springer (India) Pvt. Ltd. is part of Springer Science+Business Media (www.springer.com)

To Vinay

Preface

This book is as much about Indian philosophical traditions as it is about nature. The interweaving of both these themes is inevitable, not only because they are conceptually related but also because as a researcher I am profoundly related to both. The beginnings of this project are based on my interest in nature and in Indian philosophy. I spent a few years of my life living in the Himalayas, studying philosophy in a traditional way. During those years I experienced a deep relationship with the environment around me. A strong sense of commitment to a vision of an ethical human–nature relationship became a guiding value in my life. My struggle to find newer ways of thinking and articulating my experience led me to study psychology first and then move on to philosophy.

For my Ph.D., I wanted to study the relationship between origins of Indian philosophical thought and nature. In the very first conversation I had with my doctoral supervisor, Dr. Sundar Sarukkai about nature in Indian thought, I poured out a torrent of ideas about conserving nature, eco-ethics and people's attitude to nature. He listened patiently and then stopped me with a simple question. He asked "What is this nature you are referring to? What is nature in the Indian traditions?" What was supposed to be a short answer to this question, turned out to be the background question for my doctoral thesis. This book is a later version of my thesis that contributes to this question, "What is nature in Indian philosophy and culture and what do we do after we find that out?"

When I started out, my study was unorthodox and there was no clear disciplinary boundary. This was both a challenge and an advantage. The challenge was to keep the work descriptive and conceptual and yet not devolve into a discourse of prescriptive culturalism. The advantage was that I had an opportunity to produce a work on Indian philosophical thought and relate it to something contemporary like ecological ethics. The book reflects this struggle and balance in its various sections. In the course of my work I found that cultural geography as a discipline, in a way links the contemporary ideas to the pre-modern philosophical concepts of nature. It was also evident that certain ideas around Indian Ethics would also have to be examined and recast into the ecological ethics problematic.

Though I worked with Sanskrit texts, my approach was not that of a Sanskritist. I was familiar with traditional methods of reading Indian philosophical texts and had a working knowledge of Sanskrit. I was thus able to access the texts, using a conceptual approach. To create conceptual categories from Indian texts is to engage in an act of translation. In this method, I am deeply influenced by the work of my supervisor Dr. Sundar Sarukkai.

On the other hand, I must confess working with only some philosophical texts felt incomplete and so I included some cultural traditions such as medicine and literature to my study. I felt that this enriched the philosophical themes.

A part of the pragmatic approach to my topic comes from the tradition of environmental science and wildlife conservation, which were my earlier career paths. My co-supervisor, Dr. Anindya Sinha, regularly challenged my ethical claims and helped me make those crucial links between philosophy and conservation biology. The final two chapters of this book are a reflection of the attempt to make this work conceptually relevant to the discourses of nature conservation and place the themes within the larger context of today's ecological crisis.

Many references and earlier work from Western philosophical traditions were available on the subject. Kate Soper's *What is Nature?* was an important influence on the way I structured this book around themes. For sources from Eastern traditions, I relied on many smaller yet relevant papers by eminent philosophers working on Indian philosophy such as J.N. Mohanty, B.K. Matilal, Purushottama Bilimoria, Vasudha Narayan and others. *Nature in Asian Traditions of Thought*, an edited volume of essays by J. Baird Callicott and Roger T. Ames, gave my research a legitimate ground in the discipline of environmental philosophy.

The book also includes some chapters that were not a part of my original thesis. These chapters have been extensions of the same work undertaken in part during the spring courses on environmental philosophy that I teach my Master's students at the Manipal Centre for Philosophy and Humanities, Manipal University. The questions and discussions on this subject have led to extensive work on aspects of cultural geography and literature in this book.

While writing this book, I made a deliberate choice about not using gender neutral terms, particularly when I refer to earlier work by authors on nature. The choice of the word 'man' as against 'human' is a point to be deliberated on, particularly since a substantial part of this work focuses on eco-feminist issues. Karen Warren, whom I briefly met at a conference in the US, spoke to me about the importance of not neutralising these references which have engendered our understanding of nature and culture.

Acknowledgements

I acknowledge the unconditional love and support of my husband Vinay Baindur in my life. He has contributed in many ways by being an unshakable stand for my work. He continues to inspire me.

I am deeply indebted to my associates, friends and colleagues at Manipal University—Dr. Sundar Sarukkai, Dr. George Varghese, Dr. Nikhil Govind and Dr. Gayathri Prabhu. I thank them especially for the endless discussions and philosophical debates that crystallised my concepts and helped to make connections in my research. They bravely encountered and responded to my random philosophical thoughts at all times. I am deeply grateful to my students, Ivan, Asim and Kalpita, with whom I had many conversations that inspired questions and ideas for the book. Madhava and Tapaswi have been my Sanskrit companions and provided valuable time to discuss texts and translations. Heartfelt thanks are also due to Dhanwanti Nayak, who was always available with her advice and support. I thank Sharmada, who untiringly went through the chapters and proofread the final copy. I specially thank the staff at MCPH, Ms. Swathi and Ms. Jayanthi, for their support.

I thank Dr. Sundar Sarukkai, my guide and Dr. Anindya Sinha, my co-guide, whose intervention changed the course of my life and turned it towards a lifelong passion for research. They have been a source of intellectual stimulation and inspiration as I went through the ups and downs of a doctoral programme. As teachers and my toughest critics, their sense of humour and rational dissidence has created rigour and fresh perspectives in my study. I deeply acknowledge Dr. Sundar Sarukkai's contribution in particular for the various discussions we had on philosophy. I thank him for introducing me to the world of philosophy of science and the practice of thinking, writing and for encouraging me to express myself. I could not have had a better mentor or teacher. I thank Dr. Anindya Sinha for being available and for responding patiently to my endless questions and arguments. By challenging me to critically examine so many of my typical assumptions about conservation and ecology, he rekindled my interest in the biological sciences.

Among the faculty at the National Institute of Advanced Studies, Bangalore, Dr. Carol Upadhya and Dr. Rajesh Kasturirangan, the additional members on my doctoral committee, have always been available for substantive advice and for guidance in the interdisciplinary areas of my research and I am deeply grateful to them. Dr. Carol Upadhya introduced me to the fundamentals of anthropology and sociology that have influenced my work deeply. Dr. M.G. Narasimhan was a constant source of support and pointed me to many of the books that greatly helped me with my work. I am grateful to Dr. Purushottama Bilimoria for his encouragement and guidance on important aspects of my work. Dr. Raghuramaraju, one of the external examiners of my thesis, was insistent that my work should be published as a book. His comments on my thesis helped me to greatly improve my ideas while converting my thesis into a book. Dr. Gopal Guru who shared his views on the issues of social justice and environment inspired me to take on speculating about some applications of my research in the real world.

My colleagues Sowjanya Peddi and Sreeja K.G. collaborated with me on a project on urban lakes that contributed to the later chapters of this book. I am deeply grateful for the interdisciplinary skills I learnt while working with them.

A large part of my philosophical training was gained from learning the scriptures from my mother Narayana Lakshmi, who being a translator and a philosopher herself was able to tide me over many difficult passages. My brothers, Naren and Vivek were my philosophical companions and also help in editing earlier versions of my thesis. As an inheritor of the oral traditions of studying philosophical texts, I must confess that the insights from a long lineage of teachers and preceptors have also contributed to this work. This tradition endures within a particular notion of knowledge that does not lend itself to modern ways of author–date citations. Instead, what can be acknowledged on my part is a deep sense of gratitude, a humility and acceptance of receiving the priceless gift of knowledge. I take this opportunity to thank the line of philosophers and teachers in Indian traditions who have contributed to my thinking and reflections in very many ways.

Contents

About the Author

Dr. Meera Baindur is currently faculty at Manipal Centre for Philosophy and Humanities, Manipal University. She has a doctoral degree from Manipal University in the interdisciplinary area of environmental philosophy which she completed through the National Institute of Advanced Studies, Bangalore. Earlier, she stayed with village communities in the Himalayas for a few years, working directly on environment and sustainability issues. During this time she also pursued traditional studies in Indian philosophy and yoga. Her research interests include environmental philosophy and environmental humanities: conceptualisation of nature in Indian thought, ecological ethics and environmental ethics. Her recent publications include papers related to the idea of place. She has taught and lectured on traditional Indian philosophy, philosophy of religion, Hinduism, mythology and narratives. Her recent work centres on religions and ecological practice, rituals and cultural geography.

Chapter 1
World Views and Issues Around Nature

Abstract This chapter explains the background to the conceptualisations of nature. It starts with a general historical overview of world views of different ancient civilisations about nature. The introduction then lists the philosophical issues and themes that are prominently in discussion around the concept of nature and the relationship of human beings to the non-human world. These themes and issues will form the context for further discussions in the succeeding chapters.

Keywords Nature · World views · Conceptualisation of nature · *Physis* · *Tzu-jan* · Woman and nature

1.1 Introduction: Diverse World Views of Nature

The world we know of and inhabit as living beings is our planet Earth that we sometimes call our home. The Earth with its resources such as air, water, and soil sustains us and other forms of life. Many writers have remarked that life as it occurs on this planet is a rare and wondrous thing. Sages in ancient India have even composed hymns in praise of this life force and said that it is the essence of all creation. While all of us acknowledge and recognise the uniqueness of ourselves and the cohabitants of our world, as of now, we seem to be strangely neglectful of the very resources that sustain us. The demands that we impose upon our planet are causing the depletion, pollution, and wasteful destruction of habitats, water, and air. In a short period of our human history, as a species we have transformed our habitat in many irreversible ways. When we ask the question as to what is it that sustains us and what are we are dependent on, the sages in pre-modern India would have said "life". Instead, today we would easily use either of these two words that are closely related—nature and environment—to talk of our dependency. Not only are we aware that we depend on nature, we are also clear that we are altering and changing it in many ways. We talk of destroying nature, polluting our environment, or causing an ecological crisis to refer to the state of crisis we, as humanity, believe

M. Baindur, *Nature in Indian Philosophy and Cultural Traditions*,
Sophia Studies in Cross-cultural Philosophy of Traditions and Cultures 12,
DOI 10.1007/978-81-322-2358-0_1

ourselves to be in. Though "nature" and "environment" are used interchangeably in common parlance, specifically the term environment is associated with a somewhat functional definition. It is our surrounding that provides us conditions for growth and development and also is a source of danger and destruction. Sometimes, the environment (in ecological sciences) is described by scientists as consisting of both biotic and abiotic components of our planet.

The second term "nature", however, is not that easily defined or interpreted. This term is linked to many interpretations and discourses—culturally, politically, and scientifically. The study of the concept of nature has become the foundational enquiry of a relatively new discipline of philosophy called environmental philosophy. The scope of environmental philosophy is to understand the relationship between human beings and their environment from different perspectives including the belief and knowledge that communities hold about nature, their ethical, and moral interest towards nature and the various pre-suppositions that exist in such perspectives.

The history of the idea of nature and the history of our relationship to it have been influenced by a number of sociological, historical, cultural, and political factors. There is also a growing interest in studying these contexts to understand the implication of such influences on human attitude and behaviour towards nature. Brennan and Lo (2010) write:

> A great deal of work has theorised the pathology of contemporary environmental crises suggesting that some of our underlying cultural, religious and political beliefs and attitudes are responsible for our behaving badly towards the environment. In other words, our religious worldviews, our basic political and social ideas, are not environmentally innocent (p. 7).

One could also say that these conceptualisations are not only influenced by the sociocultural contexts of various communities of people, but one can also posit that they would also vary according to the language sensibilities of these different communities. Various thinkers have attempted to trace the history of the idea of nature (Collingwood 1945) and the conceptualisations and discourses around nature (Soper 1995) and its position within the framework of political ecology (Latour 2004). Many overviews of conceptualisations of nature that have been published as edited papers (Callicott and Ames 1989; Marshall 1992; Leiss 1994) deal with specific perspectives, regional interpretations, and periods of history.

A brief review of literature in environmental philosophy brings to our attention the various meanings and interpretations of nature, during different periods of time even within the same culture. What this survey indicates is that the idea of nature was linked closely to what one may call a "world view" of these civilisations. The world views were inclusive of idea of nature in a holistic sense and these conceptualisations predate our current understanding of nature.

Brunn and Kalland (1995) point out that concepts such as nature and others are powerful, multifaceted abstractions. Conceptualisation of nature straddles the objective world we inhabit as well as our subjective experience of it. They suggest "The qualities of our environment conceived by such concepts, are products of human consciousness as much as they are universal and objective" (p. 9). To unpack the nuances of the concept of nature and explore its extensions and implications is significant, particularly in non-Western traditions of thought. However, one would also have to contextualise such conceptualisations within the larger framework of existing interpretations of nature and also within the contemporary eco-ethical issues such as conservation and sustainability. We could begin this exploration by first examining some earlier understandings of nature and the environment in pre-modern traditions.

One finds that traditional philosophies of nature were concerned with the metaphysics of nature and humanity, cosmological narratives, and theories of the creation of the natural world. Besides rich descriptions of the natural world and its composition and components, many of these traditions included conceptualisations of nature that were related to particular historical, religious, and cultural contexts, within which such ideas emerged. We find that many of the later developments of conceptualisations and nature find their roots in some of these ancient pre-modern traditions.

1.2 Early Greek Views

Early Greek philosophers such as those of the Myelasian School and others subscribed to the view that "matter" and "life" are inseparable. This view is referred to as the hylozoic view. The Greeks believed that there are natural things that are self-produced, as opposed to things produced by skill, and that all of these natural things constitute a single "world of nature." These things could be collectively spoken of, since they shared other qualities besides naturalness. One such important property of these natural things was that they were all made of a single substance or material.

Copleston (1962, 1993, p. 21) calls the philosophy of this period as "instances of abstract materialism". He writes that the world, according to most of these thinkers, was made of an essential, imperishable stuff. He also suggests that these early philosophers assigned the idea of a primitive element being "the stuff" of the universe, though they were not materialists. This stuff was not just limited to the physical matter, but was also the stuff of thought and souls. It was substance and power, and it was also divine. It lacked any particular quality such as solidity, liquidity, or gaseousness. In this cosmos, which was made of this one-stuff, man and nature were in a continuous hierarchical order. The Pythagorean school was pantheistic in its basic outlook. Since they believed in the transmigration of souls not only across the human body but also across animals, they abstained from eating meat.

If that is so, then all life is akin, and the kinship of nature is another Pythagorean tenet. It went further than we might think, for the animate world extended further for them than it does for us. They believed indeed that the universe as a whole was a living creature (Guthrie 1950, p. 34–35). The term physis, translated by Romans into the word natura, represented a conceptualisation of "origin and growth of the universe in its entirety". The concept of nature was "physis" that existed in contrast to "psyche"—the life principle. Pojman (2003) suggests that a change did not occur in Greek thought:

> But this narrowing of physis in the direction of "physics" did not occur in the way that we imagine today. We oppose the psychic, the animated, the living, to the "physical." But for the Greeks all this belonged to physis and continued to do so even after Aristotle. They contrasted it with what they called thesis, thesis, ordinance, or nomos, law, rule in the sense of ethos (p. 1192).

Later, the same principle is contrasted with the artificially made human products: "The meaning of physis is further restricted by contrast with techneē—which denotes neither art nor technology but a knowledge, the ability to plan and organize freely, to master institutions (cf. Plato's Phaedrus). Techneē is creating, building in the sense of a deliberate producing" (p. 1192).

Another significant point is that nature as physis was never equated directly to the earth or the earth—goddess Gaia. McClure (1933, p. 120) confirms this: "No Greek ever identified physis with the earth." However, the name of this ancient goddess is now used in a specific movement of ecological thought that originated from the recognition that the earth is one of the rarest planets we know that can sustain and support life. A resurrection of the concept of the Gaia as a living organism (often called the Gaia hypothesis) has resulted in strong appeal for environmental ethical behaviour. Lovelock (1982) explains the main ideas of his hypothesis thus:

> We have since defined Gaia as a complex entity involving the Earth's biosphere, atmosphere, oceans, and soil; the totality constituting a feedback or cybernetic system which seeks an optimal physical and chemical environment for life on this planet. The maintenance of relatively constant conditions by active control may be conveniently described by the term 'homoeostasis.' (p.10)

Since the Greek tradition bequeathed its conceptualisations to Western traditions of thought, further descriptions will be surveyed in Chap. 2.

1.3 Early Indian Views

In the archaeological evidence found in the Indus valley excavations, a seal portrays a seated figure surrounded by many animals. Many of these seals portray only animals such as the bull or the elephant, and trees and leaf motifs. Thapar (2002, p. 84) suggests that these seals could have been used for stamping packages or as identification of merchants or supervisory managers. However, the representation of non-human elements in the seals of one of the oldest river valley civilisation draws our attention to an understanding of a world that is not exclusively human.

Some of the earliest views of nature in Indian thought are found recorded in textual traditions called the Vedas. Beginning with the worship of nature gods and reverence hymns to the earth, a large part of the belief system of these earliest thinkers was deeply influenced by close observation and contact with nature. Marshall (1992, p. 2) refers to the Vedas as having "a life-affirming and earthy naturalism" parallel to their deeply spiritual messages. He writes "At the same time, the Ṛgveda ('Songs of Knowledge') reveal an intimate bond between the believer and his or her environment, a sense of kinship with the spirit that dwells in all things". Macdonell (1927, p. 2) describes the phenomenon of nature gods in Vedic mythology: "The foundation, on which Vedic mythology rests, is still the belief, surviving from a remote antiquity, that all objects and phenomenon of nature with which man is surrounded, are animate and divine". The Vedic hymns also certainly demonstrate a love for the beauty of the environment. A number of metaphors and descriptions of the natural world are linked to eulogies. Some rituals described in the Vedas are invoked in many traditions of thought and persist in rituals and practices.

One of the most noteworthy ideas around the concept of nature found in the world views of the Vedic period is that of the substantive oneness of all creation, described first in the spiritual sections of the Vedas called the Upaniśads. With the development of Hinduism, we find various traditions of thought continuing to engage with nature, both spiritually and morally. Many philosophical schools in India that engaged with the concept of nature in later period were influenced by the ideas of the Vedic period. Some of them will be discussed in detail in the forthcoming chapters.

1.4 African Views

As with the case of Indian thought, it is certainly unfair to club the diverse streams of African thoughts into a monolithic "African view of nature". However, African scholars have suggested that it is possible to trace some common ideas among the various tribes, peoples, and communities of the region. Although most communities believe in a God, the African view of nature is described as an inclusive notion, with the tribe, person, and land forming a single unit. Nobles (2000, pp. 280–281) notes that the two fundamental operational orders among the different tribes of Africa are the ideas of being one with nature and the notion of survival of one's people. He writes "Hence the African experience defines man's place (role) in nature's scheme" (p. 281). Callicott (1994) unpacks the understanding of this "bio-communitarianism" further:

> An African's identity, nevertheless, is not confined to his or her role in the community. African social psychology is not modelled on the anthill, the beehive, or the termite colony. Each individual is a distinct person, with his or her unique blend of personality, needs, desires, talents, and destiny. But, far more vividly than in the modern Western worldview, individuality is not only counterbalanced by community identity but one's unique individuality is defined in part by one's social relationships and expressed through social interaction (p. 166).

1.5 Chinese Views

The Chinese concept of nature can be understood from the ancient philosophy of Taoism prevalent towards sixth century BC. This philosophy stressed the idea of living in harmony with nature. In contrast was the Confucian world view that promoted a strict hierarchy of nature and human beings. Both these philosophies drew their inspiration from the principle of *Tao*, sometimes called the "natural way", which was conceptualised as a flowing and creative process consisting of the two complementary forces, *Yin* and *Yang*.

Marshall (1992, p. 9) contrasts the two schools and their different perspectives: "The Confucians celebrate traditionally 'male' virtues like duty, discipline and obedience, while Taoists promote the 'female' values of receptivity and passivity". He points out that the "former wish to dominate and regulate nature; the latter to follow and harmonize with it". Marshall (1992, p. 9) also points out that this struggle between the two world views is still with us. According to him, the Chinese phrase that is translated as nature is *tzu-jan* (also written as *tsu-jan*) and this refers to the idea of spontaneity, a creation that happens by itself without the creator. This stresses upon the perspective that nature is self-created. In Lao-Tse, the following passage speaks of *tsu-jan*: "Man is based on earth, the earth is based on heaven, heaven is based on the Way (Tao) and the Way is based on nature (*tsu-jan*)" (Lao-Tse, quoted in Tellenbach and Kimura 1989, p. 153). The school of Taoism had a view of nature that required one to relate to nature without pretences (Parkes 1989): "The tao way of relating to nature is to see through and withdraw the ego-generated anthropomorphic projects that vitiate our relations with natural phenomenon, allowing us to live in what Chuang-tzu calls 'a full view of heaven [T'ien]" (p. 80).

According to Taoist thought, the world is a sphere divided into sections represented by *Yin* and *Yang*, which is described as oriented in particular directions.

> The world is a circle (or a sphere) divided along two planes. The vertical cleavage goes from north (seen as down below) to the south (up above): on the left-hand side is Yang, on the right-hand, Yin. The other cleavage is horizontal and marks off an upper half, which is Yang, from a lower half, which is Yin. As a result, the sphere is divided into four parts: the southeast is Yang, and the northwest is Yin the other two parts are made up of Yin and Yang combined, with the northeast sector, according to the Chinese, that of Yang in Yin, and the southwest that of Yin in Yang (Robinet 1997, p. 9).

According to Robinet (1997), *Yin* and *Yang* are not mystical male and female energies as is popularly understood. They are "lines of force", interacting with each other in complex and contradictory ways. She clarifies

> We might well say that Yin and Yang are two extreme poles, two ideals with only a conceptual and didactic existence. They do not exist in the world, but all things tend toward them to one degree or another. They govern a liturgical division of the universe and its double generative process. They are the basis of the difference that gives rise to attraction, as well as of all development and the multiplicity produced through their combinations (p. 9).

Taoism places the human at the centre of the cosmos and urges the formations of interconnections that will lead to a relationship between the aspirant and the cosmos, a rediscovery of what they call the prime mover: "… the purpose of Taoism is to create a clear and close union between the interior and exterior world. The exterior world is understood above all as consisting of Nature, the cosmos, the natural world, and only secondarily of human society" (p. 18).

Chinese neo-Confucianism is regarded by Tucker (1991, p. 62) as having a cosmological sensibility that is characterised by holism and dynamism. The microcosm and the macrocosm coconstitute the unified nature which is seen as interrelated. Many texts such as the *I-Ching* elaborate on the different aspects of these correspondences and correlations (Tucker 1991). She explains "Within the context of correspondences the relation of oneself as microcosm to the universe as macrocosm is a central theme arising directly from the underlying idea of organic wholeness" (p. 63). According to Wei-ming (1989), the Chinese conceptualise their cosmos as composed of *ch'i* that can be translated as matter-energy (p. 71). He adds that the universe is viewed as a chain of being, an "all-enfolding harmony": "It means that nature is all-inclusive, the spontaneously self-generating life process which excludes nothing." The word *tzu-zan*, used in modern Chinese captures this meaning, claims Wei-ming (1989, p. 71). Such transformation of the universe is not linear. In this conceptualisation, nature that is ever-expanding "exhibits three basic motifs: continuity, wholeness, and dynamism". Unlike the Taoist thought, the world is not a sphere. Ming suggests that Chinese cosmogony is an open system that can be referred to as spiral or cylindrical but as transformational. He claims that the ever-expanding universe of transformation does not lend itself to geometric metaphors (p. 71).

1.6 Japanese Views

According to Shaner, there are three distinct traditions of philosophy in Japan that have had influence culturally: (1) Buddhist philosophy (from Korea and China); (2) Confucian philosophy (from China); and (3) Shintō (which is indigenous). The Japanese term that corresponds to nature is "shi-zen" or "*ji-nen*" (Tellenbach and Kimura 1989). The term is derived from the Chinese term *tzu-jan* and almost corresponds to its meaning in many ways. Hubertus Tellenbach and Bin Kimura (1989) point out that before this term was used in Japanese, there seemed to be no single concept of nature that was inclusive of all natural things. They suggest that there was a term *onozukara* which represents "an objective state which begins of itself without any external mediation" (p. 153). They explain that this in contrast to the subjective state that someone carries into completion, a state of human contrivance called *mizukara* (p. 156). According to them, *shi-zen* has an original Japanese meaning that "… names something like 'mountain, river, ocean, plant, animal, rain, wind etc.,' signifying therefore a way of being, which exists without human intervention" (p. 157).

The indigenous tradition of Japan encouraged an attitude to nature that did not seek to read meanings beyond the simple and direct perception of nature. This "non-symbolic attitude" to nature was 'natural' in the sense of its purity, sincerity, and simplicity (Tellenbach and Kimura 1989). Nature thus had intrinsic value within this world view.

We find that in all of the above discussions, there are two themes that may be of interest to us. One recurrent theme is the idea of the celestial bodies, stars, and the planet Earth and the description of the creation of the cosmos that figure predominantly in narratives and explanations of these world views. In other words, the attempt of these ancient scholars is to give an explanation of the process of creation, provide an exhaustive list of the various elements of the cosmos, and describe how they are subject to a natural order or control. The second theme that may be of greater interest to philosophers of nature is the place of the human being within the grander scheme of the cosmos. We find that many of pre-modern philosophies also laid great emphasis on describing the relationships between elements and beings, while ordering and classifying them. We shall see in Chap. 2 how some of these attempts were to have a great impact on the idea of nature.

At this point, a note on the objectives of this book may help the reader find coherence in the various chapters and themes within. The book conceptually shifts between two themes. Firstly, there is an attempt to distinguish the conceptualisation of nature in Indian traditions of thought, largely through selected texts from the body of philosophical literature available in Sanskrit. A few references to other traditions that are oral or landscape-related traditions will also be taken up for discussion as they are significant to some themes and cannot be ignored. Secondly, there is an attempt to engage in some speculations about how these conceptualisations would make sense in contemporary environmental context. There is of course a possibility that such speculations may not have any direct implication for the ecological crisis, but it is hoped that they will stimulate newer and nuanced ways of thinking about nature. Such understanding may also indicate future directions for conceptual change and ideas around eco-ethical action. Further, pragmatic interventions can be then derived from eco-ethics, informed by alternative conceptualisations. There is of course in such a project, an implicit philosophical objective, that of resolving philosophical and conceptual issues within environmental philosophy itself. Before we go on to explore the idea of nature in some detail, we must explore some of the key issues around the conceptualisation of nature that are of interest to us today. These questions may form a framework or a background for us to understand how one could discuss conceptualisation of nature in Indian traditions of thought. Understanding some of these issues will also enable us to foreground some debates around the concept of nature.

1.7 Issues Around the Concept of Nature

We have seen very briefly the diverse interpretations and meanings of nature in various civilisations of the past. Today, however, we find that most (if not all) of these types of conceptualisations are no longer prevalent in the way our modern world thinks and discusses nature. Environmental philosophers in the early twentieth century began to examine the complexities of the relationships between human beings and nature. This area of study has gained importance, given the current state of the world where there seems to be more and more evidence every day to confirm that human action is affecting the environment in complex ways that are both unprecedented and unsustainable. On one hand, there seems to be an attempt to interpret nature as a uniform, measurable, and understandable entity on the part of the various sciences. On the other hand, many environmental philosophers agree that the current view of nature has various shortcomings that do not admit perspectives on new ecological challenges facing us today (Larson 1987). The subsequent section is an attempt to unpack and describe some of those issues around nature that are significant to the cause of ecological thinking in particular.

Within the wide range of environmentalist concerns about the impact of human beings on nature and the environment, it is clear that many ideological, philosophical, and social pre-suppositions come into play. We find that most of the discourse on nature has focused on nature as being representational. Any representation then is contained within a set of pre-suppositions and cultural values that are held by the people making representations. Fundamental to the study of these representations is the conceptualisation of "nature". It can be posited that fundamental world views about nature, especially those pre-suppositions that have had a direct effect on the way human beings relate to nature, are given by particular conceptualisations of nature. In the field of conservation and conservative action for instance, phrases such as "protect nature" or "nature conservation" are commonly used. Hidden within these phrases is an idea that one can recognise nature and identify it when we see it and make uncomplicated choices between natural things that are to be protected and unnatural things that are not nature. The question is can we really make such uncomplicated choices? Though the direct connection between conceptual resources and actual practices cannot be proved, yet it is clear that a series of conceptual elements form a background to any world view, including the ecological and these do influence people's attitudes.

Prevalent literature in this area lays claim to the opinion that world views form an important framework for looking at human relationship to nature. The notion of "world views" carries within it a sense of historical, cultural, and geographical framework that goes far beyond the traditional engagements of philosophy. Even though pre-suppositions about the natural world may not affect behaviour, associative values impact a large number of individual and community decisions about nature. Lafreniere (1985) insists that "the majority of individuals living in any culture are motivated by a particular world view". He argues that, to effectively create an environmental ethics education programme, it is important to study and understand world views. Such an understanding is important in order to deal with the causes of the environmental problems which are cultural, rather than deal with the symptoms that are evident in the natural world, he notes.

1.8 Definition of Nature: Ambiguities and Interpretations

If nature only represented the world of plants, trees, and animals around us, why would there be any ambiguity? We are often so sure of its meaning when we use the word nature in our daily conversations. But if we further ask ourselves what exactly nature is, or ask if the fly on our kitchen wall is a part of nature, we have to pause and really think about what we are referring to when we say "nature". This ambiguity around conceptualisations is because "nature" is much more than a mere descriptive word for the physical reality of our surroundings, or what we would call the natural world. Many philosophers are of the opinion that the idea of nature has been subject to various themes of reduction—such as naturalism or Cartesian dualism—and also to the idea of a static nature that is the object of conservation and management programmes.

Even when used in everyday language, the term nature has many layers of meaning given by its historical and cultural contexts of the past. Soper (1995, p. 9) refers to these various layers as a "symbolic load", that the term carries. Most of this "symbolic load" has been gathered through a long historical process of conceptualisations and usage of the term by various communities and cultures particularly in Western philosophical thought. The interpretation of nature gets further complicated within newer contexts such as conservation efforts or within the frameworks of newer disciplines such as ecology or conservation biology.

As we search for the first use of the word nature, we find the word *natura* in Latin that gave us the term nature. According to Williams (1989), who explains it clearly, "nature" comes from the word *natura* in Latin, the past particle of "*nasci*", the root word which means "to be born" (same as the root of "native", "innate", etc.). Its earliest meanings are also associated with the essential quality of something (sounds vague). However, as we have noted earlier, the term "*physis*" was translated by Romans into the word "*natura*". "*Physis*" in Greek thought particularly was associated with growth.

However, though there were many possibilities of the study of nature among the different schools of Greek thought, Western philosophy seemed to move away from the more organic and active meanings of nature. Callicott and Ames (1989) also point out that the one tradition that lent itself to the development of a particular system of knowledge that we now know as science was that of natural philosophy:

> The first Greek philosophy was natural philosophy, and although many ecologically adaptable or environmentally useful ideas were broached, the natural philosophy that was culturally selected by this dialectic of Western intellectual history, and thus survived to bequeath its characteristics to the modern period is atomism (p. 5).

Many issues surrounding the idea of nature are related to this one dominant understanding of nature that developed from this particular school of thought. However, it is possible that this reductive understanding of nature is no longer sufficient to deal with the complex questions of an ecological crisis. Nature, in its simplest literal meaning, refers to the non-human components of the universe. But, this is a dominant view derived from Western traditions of thought. On the other hand, there are scholars who question the very reality of "nature" itself. In the next section, we explore one of the issues that have arisen out of the interpretation of "nature".

1.9 Place of Humans in Nature: Inclusive and Exclusive Views

The history of conceptualisation of nature is dependent on the sociocultural narratives and discourses of societies, making a strong case for the social construction of nature. Under this category of understanding nature as something culturally interpreted, different ideas about human relationship to nature is also proposed. Some people believe that human beings are included in a "nature" that refers to the natural world at large, while others believe that human actions and the impact on the planet actually exclude the human beings from the rest of nature (as illustrated by the commonly used phrase "man and nature").

The idea that human beings are not like other beings by virtue of their rationality and creative powers can be traced to early Greek thought. The Greeks also believed that there are artificial objects produced by human contrivances which are very different from natural kinds of things.

White (1967), whose essay is regarded as one of the first critiques of a Western view of nature, helped to initiate debates in the area of a largely human-centred thinking of Western religions and suggests a more nuanced, egalitarian view of nature. He writes:

> Especially in its western form, Christianity is the most anthropocentric religion the world has seen.... Christianity in absolute contrast to ancient paganism and Asia's religions [except, perhaps, Zoroastrianism], not only established a duality of man and nature but also insisted that it is God's will that man exploit nature for is proper ends (p. 1205).

Once there is the predominance of a non-inclusive view of nature or a view under which somehow the human beings and their value is given more prominence, we find that the binary of humans and non-human nature becomes a duality that has deeper implications for our ethical stance towards nature. Soper (1995) argues that the demarcation between humans and nature is fundamental to any prescription in ecology or conservation. In other words, in the debate between Humanism and Ecology, the questions of "what is nature?" and "where is the place of human being in nature?" cannot be ignored. Environmental philosophers believe that a conceptualisation of nature which does not distinguish between the human and the non-human world would create a better framework of environmental ethics. For instance, the first United Nations' World Conservation Strategy, published in 1980, has this to state:

> Ultimately the behaviour of entire societies towards the biosphere must be transformed if the achievement of conservation objectives is to be assured. A new ethic, embracing plants and animals as well as people, is required for human societies to live in harmony with the natural world on which they depend for survival and well-being. The long term task of environmental education is to foster or reinforce attitudes and behaviour compatible with this new ethic (IUCN report, 1980).

As a discipline, environmental ethics is very recent, but human concern for the environment and nature in the form of practices has always existed in the past. The field of ethics, also called moral philosophy, involves systematising, defending, and recommending concepts of right and wrong behaviour. Ethical frameworks, which have traditionally dealt with the behaviour of human beings towards other human beings or at the most towards other living beings, are now applied in the immense task of research in environmental ethics. In the literature on environmental ethics, the distinction between instrumental value and intrinsic value (i.e. non-instrumental value) has been of considerable importance.

The idea that nature has an intrinsic value apart from the utilitarian values it has for the human beings is a significant idea that forms the basis for what is called "bio-centric" view of nature. Some philosophers like Taylor (1986) believe that such a conceptualisation of nature would do away with the exploitative paradigm with which human beings relate to nature. Brunn and Kalland (1995) indicate that moral obligations in Asian cultures are based on continuity of the human and the non-human:

> Nature and morality are closely linked in many Asian cultures, man and environment forming a moral unity. Yet as there is no absolute good or evil, there is no absolute morality, at least not for commoners (p. 11).

One value that has had a great impact on Indian thought even in the current age is that of non-violence/non-injury (*ahiṃsa*). This value comes to us from the traditions of Sāṃkhya-Yoga and Jainism. Many scholars appreciate the overarching moral considerability of *ahiṃsa* and also acknowledge other forms of empathetic beliefs in later movements of Hinduism such as *Bhakti* (devotional) traditions. Others argue that it is not really required to grant intrinsic value to nature to create a framework for environmental ethics. These philosophers believe that as human beings who are separate from nature and use nature for our own survival, we would still care for the environment because ultimately, human welfare depends on the environment. The argument is that it is unnecessary to invoke the idea of an intrinsic value of nature for conservation. This view is referred to as "anthropocentric".

1.10 Nature as Pristine Wilderness

Henry David Thoreau (2001) says in his essay *Walking*, "In Wildness is the preservation of the World". Wilderness is a problematic concept in environmental philosophy that has at its root the distinction between what can be called natural versus what has been worked with by human beings, the unnatural or modified. Wilderness particularly refers to landscapes that are exclusive of human interference. Lewis (2007) also distinguishes between the quality of wildness and the concept of wilderness in terms of the scale of the landscape areas that require to be designated as wilderness:

> Wilderness is a concept devised by humans to define a particular type of wild environment —with its plants, animals, and ecosystems—and it is entirely appropriate to declare that wilderness, as distinct from wildness, must be large on a human scale. Wild nature can be found everywhere; wilderness cannot (p. 6).

The popularity of the idea of the wilderness being invoked for conservation is based in a particular history of ideas in Western thought, particularly in North America. Nelson (2008) remarks:

> Wilderness, unlike many of the words for the things within a Wilderness area, is not readily translatable into a wide variety of languages. This linguistic lack forms, in part, the first critique of the received view-that it is ethnocentric (emanating from one culture and inappropriately applied to other cultures) (p. 201).

Callicott and Nelson (1998) suggest that wilderness is received idea and not a natural category of understanding nature. Critiques of wilderness and debates around the reality of wilderness are prevalent and much discussed. Cronon (1995) suggests that the problem of wilderness is embedded in two broad conceptual transformations, the romanticism and the frontier imaginations both of which culturally construct wilderness as value-laden. He writes:

> The two converged to remake wilderness in their own image, freighting it with moral values and cultural symbols that it carries to this day. Indeed, it is not too much to say that the modern environmental movement is itself a grandchild of romanticism and post-frontier ideology, which is why it is no accident that so much environmentalist discourse takes its bearings from the wilderness these intellectual movements helped create (pp. 474–475, in Callicott and Nelson 1998).

He points out that the cultural construction of nature as pristine, imbues it with an almost sacred quality. Landscape experiences were not just pleasurable experiences but began to be articulated as divine experiences. The landscape experience is likened to the experience of divinity, beyond aesthetic value and the joy of being in nature. The second cultural movement that impacted the idea of wilderness was the frontier imagination. According to Cronon (1995), the return to primitivism supported by the nostalgic connections to pioneer history of the Americans lead to celebration of the wild and a call for its preservation. He writes:

> This nostalgia for a passing frontier way of life inevitably implied ambivalence, if not downright hostility, toward modernity and all that it represented. If one saw the wild lands of the frontier as freer, truer, and more natural than other, more modern places, then one

was also inclined to see the cities and factories of urban-industrial civilization as confining, false, and artificial (p. 480, in Callicott and Nelson 1998).

Current debate in wilderness centres around concerned environmental philosophers who want to make an appeal for conservation of non-humanised nature while others disagree pointing out that such a conceptualisation in the long run may lead to further degradation of areas "humanised nature" creating a value hierarchy of sorts. Nelson (2008) suggests that conceptual analysis of the idea of wilderness may help to further clarity on this.

Just as he has suggested that this is a concept exclusive to Western traditions of thought, there is no exact equivalent of wilderness in Indian traditions of thought. Conceptually, the jungle (*Jangal* in Hindi) has been associated with wilderness in Indian thought. However, this idea of wild has more to do with the sense of wild and fearful than the aesthetic sublime category it represents for debates within Western thought. In a later chapter, I explore the binary of the forest and settlement that seems to underlie some discourses of nature in Indian thought. Another category of landscape associated with beauty and function finds its expression in ancient Tamil poetical works that will be discussed briefly.

1.11 Nature as Feminine

The abstraction of nature as a feminine goddess, or personification of a singular feminine deity such as Mother Nature or earth goddess seems common to many civilisations. The representation of nature as feminine has a long history of imaginations, narratives, and conceptualisations. This representation of nature is not like the usual animism or forms of personification, nor is it limited to the imagination of goddesses who controlled the forces of nature. Many ancient civilisations represented the earth as a mother (Gottlieb 2004, p. 5). Many rituals in agrarian communities, even in current times, represent the earth as a woman, whose fertility nurtures and produces crops. We know that in most cultures, a woman with no children and earth that is uncultivable are both referred to in the same way as "barren". Women and nature were thus connected as reproductive bodies, often compared to each other, and often treated in the same way. Shrirama (1966, p. 164) affirms this about the woman–nature connections in Indian thought: "The earth gives birth to various vegetation [sic]. The value of the earth depends upon its fertility. Similarly the worth of a woman lies in her ability to become the mother of sons." Merchant (1990) also argues that the concept of nature and woman is historical and social constructions where both were seen as passive and subordinate to men: "The new image of nature as a female to be controlled and dissected through experiment legitimated the exploitation of natural resources. Although the image of the nurturing earth popular in the Renaissance did not vanish, it was superseded by new controlling imagery" (p. 189). This imagery according to her is that nature willingly reveals herself to science: "From an active teacher and parent she has become a mindless submissive body" (p. 190).

Given that the values associated with women and the values associated with nature are similar, it is possible? It is possible that any change in the way nature is treated is likely to have an effect on the way women are treated and vice versa (Plumwood 1993; Warren 1996). Two images of nature were prevalent in Western traditions (to some extent in Indian traditions about women). One was that of a female nurturer and the other was that of a wild and uncontrollable nature (Merchant 1990). For instance, within the Indian tradition, on the one hand, we have the goddess as Durga, who is the protective mother; on the other, we have Kali, who is untrammelled force and energy. Merchant (1990) suggests that the emergence of the earth as living might have restricted certain activities of human exploitation in the west. She writes: "One does not readily slay a mother... As long as the earth was considered to be alive and sensitive; it could be considered a breach of human ethical behaviour to carry out destructive acts against it" (1990, p. 3).We find similar sentiments reflected in the Vedic verse:

> What, O earth, I dig out of thee, quickly shall that grow again: may I not, O pure one, pierce thy vital spot, (and) not thy heart! (AV XII. 12, Bhūmi Sūkta, 35, Trans. quoted from Panikkar 1977, p. 128)

These are some of the earliest textual references to nature as feminine in the Vedas where the earth is considered a mighty powerful goddess, the mother of all beings. Just as a mother provides for all her children, the earth provides nourishment and takes care of all beings dependent on her. In a later chapter, I shall point out how this earlier idea was replaced with other narratives and this could have led to the beginnings of the ideas of domination and mastery of nature in Indian traditions.

The feminine connection to nature goes beyond just the portrayal of nature as a woman. Nature and woman are placed as parallels, often on the other side of culture and man. What does this do the way we perceive nature or the woman? Eco-feminists claim that certain value dualisms in the Western tradition of thought associate woman and nature strongly (Warren 1996; Gaard 2003). The contention of these scholars is that "everything associated entire societies towards the biosphere must emotion, body, nature and women is regarded as inferior to that which is (historically) associated with reason, mind, culture human (i.e. male), and men" (Warren 1996, p. xii).

Merchant (1990) in her book *The Death of Nature* points out that the descriptions of nature as feminine and organic are associated with value systems. The changes in conceptualisations of nature are therefore likely to be accompanied by changes in cultural values. Any positioning of such values is likely to also in some ways impact the relationship of human beings to nature. Though eco-feminists hold diverse theoretical positions about woman and nature, one could say that in common, all of them reject the nature/culture dualism of patriarchal thought and locate animals and humans within nature (Gaard 2003). Due to the close connection between nature and women, eco-feminists assert that the environment is a feminist issue and historically the Western traditions have devalued whatever is associated with women, emotions, non-human, and the body (Gaard 2003).

Warren (1996) suggests that such devaluation occurs because of conceptual and historical causes of a "logic of domination" that is based on a framework of value dualism and value hierarchy. According to her, the devaluation occurs from conceptualising dualisms as "disjunctive pairs in which the disjuncts are seen as oppositional (rather than complementary) and exclusive (rather than as inclusive) ..." and then placing such dualisms within "value hierarchies" that she explains are "perceptions of diversity organised by a spatial Up–Down metaphor which attribute higher value (status, prestige) to that which is higher..." (p. xi).

Even in Indian traditions of thought, the connections between women and nature have led to similar devaluation and exploitation of women and nature. As pointed out earlier, we find the attribution of "motherhood" to the earth, rivers, and other natural features.

Sherma (2000) suggests that the sacralisation of the mother and the conceptualisation of nature as maternal feminine do not contribute to any significant ecological benefit. Instead, she is of the opinion that this "undermines the potential for ecological benefit". The ideal of motherhood is itself based on characteristics of negation and sacrifice:

> Though Motherhood is highly honored in Hindu kinship systems, it seems this honor is based on her self-negation, the ability to endure privations for the family, the willingness to give sustenance, no matter what the sacrifice, with no thought of her own needs. ...whether it is Bhūmi devi or a Bharatmata or a sacred grove, the expectation is that the sacred site will bless, nurture, purify, or perform any other supportive material act without any requirement for sustenance in return (p. 97).

As with the earth and woman in Indian traditions, both are considered only a receptacle for a seed. As a nurturer, woman's role is secondary, and the seed giver's role is primary. Chakravarti (1993) quotes Yalman's study, "... a fundamental principle of Hindu social organization is to a closed structure to preserve land, women and the ritual quality within it". In the context of the development of agriculture and private land holdings, the idea of legitimate motherhood and the notion of a single partner were established as foundations for patrilineal succession, she argues (p. 583).

In this section, we have seen that while the conceptualisation of nature as woman, mother, or goddess seems to suggest that there would be a possibility of respect for nature, it actually may not be so. Instead, the very act of seeing nature as feminine could lead to undervaluing and the exploitation of the earth and women for resources.

1.12 Nature as Object

Duncan (1991) claims that a systematic study of nature as an object was possible because historically, nature lost its personhood without losing the quality of "being in order". The project of science seems to be to discover the "nature as the order", including but not limited to causal laws and processes.

What is the problem with this view of nature? Merchant (1990) claims that the conceptualisation of nature as a mechanical system, after the scientific revolution granted legitimacy to human beings of "manipulating nature". She writes "Moreover, as a conceptual framework, the mechanical order has been associated with a framework of values based on power, fully compatible with the directions taken by commercial capitalism" (pp. 193–194). This framework according to her justifies the extraction, manipulation, and use of nature, but does not allow for an ethical stance towards nature.

What is interesting to note is that there has always been a reference to a nature that is uniform and singular. This uniform "nature" that is studied in the disciplines of sciences ignores the terrestrial, presentational idea in favour of a particularly abstract representational concept. Latour (2004) writes of this relationship between science and nature: "[But] this nature becomes knowable through the Intermediary of the sciences; it has been formed through the networks of instruments…" Further, he adds "For them [ecologists], science remains a mirror of the world, to the extent that one can almost always, in their literature, take the terms 'nature' and 'science' to be synonyms" (p. 4). The understanding of nature through the intermediary of sciences is almost hegemonic, often smothering other understandings as less legitimate.

Keulartz (1995) posits that the philosophy of nature and the science of nature were set in opposition to the history of nature. The views of nature that are worked out in terms of mathematical physics by the theoretical formulations of Descartes or Newton posit that nature is a system of things, events, and laws operating with mechanical necessity. In this view, nature is abstracted to a set of principles. While a philosophy of nature seeks to engage in the conceptualisation of nature from both metaphysical and ontological perspectives, a science of nature seeks to examine the concept of physical reality of the world around us.

This chapter has provided an introduction to some broad themes and issues around the relationship of human beings and nature, particularly from a perspective that is important for environmental philosophy. It is clear that many of the themes and issues discussed above are derived from certain historical developments in Western traditions of thought. An overview of the conceptualisations of nature, highlighting some important thinkers and writers in this area, is needed before we undertake the study of the concept of nature in Indian traditions of thought. Some environmental philosophers believe that any conceptualisation of nature that can actually address problems in the context of the above-mentioned philosophical problems should necessarily match a concept of ecological nature that includes human beings on the same level as other beings. The requirement that a concept of nature that needs to represent and present nature as a whole, and also needs to include the individual as well as the non-human entities, has been called to attention by Callicott and Ames (1989). They illustrate and explain that in the recent literature of environmental philosophy, there has been "an interest in finding new integrative and moral paradigms by means of which one can establish a mutually fulfilling and beneficial relationship of man to nature" (p. 21). Such a study must therefore take into account earlier work in philosophy and environmental history that traces some of the older paradigms and their development.

References

Brennan, A., & Lo, Y. S. (2010). *Understanding environmental philosophy*. Durham: Acumen.

Bruun, O., & Arne, K. (Eds.). (1995). Images of nature: An introduction. In *Asian perceptions of nature: a critical approach* (pp. 1–24). Richmond: Nordic Institute of Asia Studies, Curzon Press.

Callicott, J. B. (1994). *Earth's insights: A multicultural survey of ecological ethics from the Mediterranean Basin to the Australian Outback*. Berkeley: University of California press.

Callicott, J. B., & Nelson, M. P. (1998). Introduction. In J. B. Callicott & P. M. P. Nelson (Eds.), *The great new wilderness debate* (pp.1–20). Athens: University of Georgia Press.

Callicott, J. B., & Ames, R. T. (Eds.). (1989). Introduction. In *Nature in Asian traditions of thought*. Albany: State University of New York Press.

Chakravarti, U. (1993). Conceptualising brahmanical patriarchy in early India: gender, caste, class and state. *Economic and Political Weekly, 28*(14), 581–583.

Collingwood, R. G. (1945). *The idea of nature*. Oxford: Clarendon Press.

Copleston, S. J. F. (1993). The cradle of western thought: Ionia. In *A history of western philosophy, vol. 1, Greece and Rome*. New York: Image Books, Doubleday Publishing Group, Inc. (Reprint, 1962).

Cronon, W. (1995). The trouble with wilderness; or, getting back to the wrong nature. In W. Cronon (Ed.), *Uncommon ground: Rethinking the human place in nature,* (pp. 69–90). New York: W. W. Norton & Co.

Duncan, C. A. M. (1991). On identifying a sound environmental ethic in history: Prolegomena to any future environmental history. *Environmental History Review, 15*(2), 5–30.

Gaard, G. (Ed.). (2003). Living interconnections with animals and nature. In *Ecofeminism: Women, animals, nature*. Philadelphia: Temple University Press.

Gottelieb, R. S. (Ed.). (2004). Introduction. In *This sacred earth: Religion, nature, environment.* London: Routledge.

Guthrie, W. K. C. (1950). *The Greek philosophers: From Thales to Aristotle*. London: Methuen.

International Union for Conservation of Nature (IUCN). (1980). *World conservation strategy: Living resource conservation for sustainable development.* Prepared with cooperation of the organisations, UNEP-WWF, FAO and UNESCO. Available at https://portals.iucn.org/library/efiles/documents/WCS-004.pdf.

Keulartz, J. (1995). *The struggle for nature: a critique of radical ecology* (R. Kuitenbrouwer, Trans.). London: Routledge.

Lafrenniere, G. F. (1985). World views and environmental ethics. *Environmental Review, 9*(4), 307–322 (Special issue: Roots of ecological thought (winter)).

Larson, G. J. (1987. 'Conceptual resources' in South Asia for 'environmental ethics' or the fly is still alive and well in the bottle. *Philosophy East and West. 37*(2), 150–159.

Latour, B. (2004). *Politics of nature: How to bring the sciences into democracy* (C. Potter, Trans.). Hyderabad: Orient Longman.

Leiss, W. (1994). *The domination of nature*. Montreal: McGill-Queen's University Press. http://www.questia.com/PM.qst?a=o&d=80437339. (Stable URL).

Lewis, Michael (Ed.). (2007). *American wilderness: A new history*. New York: Oxford University Press.

Lovelock, J. (1982). *Gaia: A new look at life on Earth*. Oxford: Oxford University Press.

Macdonell, A. A. (1927). *India's past: A survey of her literatures, religions, languages and antiquities*. New Delhi: Asian Educational services.

Marshall, P. (1992). *Nature's web: Rethinking our place on Earth* (1st ed.). London: Simon and Schuster Ltd.

McClure, M. T. (1933). *The Greek conception of nature*. In *Proceedings and Addresses of the American Philosophical Association* (Vol. 7, p. 120).

Merchant, C. (1990). Preface: 1990. In *The death of nature*. New York: Harper San Francisco.

Nelson, M. (2008). The great new wilderness debate: An overview. In P. M. P. Nelson & J. B. Callicott (Eds.), *The wilderness debate rages on: Continuing the great new wilderness debate* (pp. 200–207). Athens: University of Georgia Press.

Nobles, W. W. (2000). African philosophy: Foundations for Black Psychology. In F. W. Hayes III (Ed.), *A turbulent voyage*. Lanham: Rowman and LittleField Publishers.

Panikkar, R. (1977). *The vedic experience: Mantramanjari*. New Delhi: Motilal Banarsidass Publishers.

Parkes, G. (1989). Human/Nature in Nietzsche and Taoism. In J. B. Callicott & R. T. Ames (Eds.), *Nature in Asian traditions of thought* (pp. 79–98). Albany: State University of New York.

Pojman, L. P. (2003). *Classics of philosophy* (2nd ed.). New York: Oxford University Press.

Plumwood, V. (1993). *Feminism and the mastery of nature*. London: Routledge.

Robinet, I. (1997). *Taoism: Growth of a religion* (P. Brooks, Trans.). Stanford, CA: Stanford University Press. http://www.questia.com/read/5710955. (Stable URL).

Sherma, R. D. G. (2000). Sacred immanence: Reflections of ecofeminism in hindu tantra. In L. E. Nelson (Ed.), *Purifying the earthly body of god: Religion and ecology in Hindu India* (pp. 89–133). New Delhi: D.K Printworld (P) Ltd.

Shrirama, I. (1966). Correspondence between woman and nature in Indian thought. *Philosophy East and West, 16*(3/4), 164.

Soper, K. (1995). *What is Nature? Culture, politics and the non-human*. Oxford: Blackwell Publishers Ltd.

Taylor, P. (1986). W. *Respect for nature: A theory of environmental ethics*. Princeton, New Jersey: Princeton University Press.

Tellenbach, H., & Bin K. (1989). The Japanese concept of nature. In J. B. Callicott & R. T. Ames (Eds.), *Nature in Asian traditions of thought*. Albany: State University of New York.

Thoreau, H. D., & Dillman, R. (Eds.). (2001). The major essays of Henry David Thoreau. Albany, NY: Whitston.

Tucker, M. E. (1991). The relevance of Chinese Neo-confucianism for the reverence of nature. *Environmental History Review*, 55–69.

Thapar, R. (2002). *The Penguin history of early India: From the origins to AD 1300*. New Delhi: Penguin Books.

Warren, K. J. (Ed.). (1996). Ecological feminist Philosophies: An overview of the issues. In *Ecological feminist philosophies*. Bloomington: Indiana University Press.

Wei-ming, T. (1989). 'The continuity of being: Chinese visions of nature. In J. B. Callicott & R. T. Ames (Eds.), *Nature in Asian traditions of thought* (pp. 67–78). Albany: State University of New York.

White, L, Jr. (1967). Historic roots of our ecological crisis. *Science, 155*, 1203–1207.

Williams, R. (1989). Nature. In *Keywords: A vocabulary of culture and society*. London: Fontana. (Reprint, 1975).

Chapter 2
Conceptualisations of Nature: The Narratives So Far

Abstract This chapter traces the history of the idea of nature in Western traditions of thought, giving the reader a background into the complexity of the idea. Earlier work by scholars is summarised with explanations and comments in order to indicate how the concept of nature has always been conceptualised through particular historical and cultural perspectives. The need to study the concept of nature within alternative traditions of thought is discussed. Building on the themes from the earlier chapter, a brief outline of some of the major disciplines linked to the field of Environmental Philosophy is also described.

Keywords Conceptualisation of nature in Western traditions of thought · Science and nature · Bacon's project · Enlightenment · Romanticism

2.1 Introduction

An overview of the conceptualisation of nature both from Eastern and Western streams of thought is presented this chapter. The historical trajectory of nature in Western thought is mapped through the writings of various historians of nature and environmental philosophers. The attempt has been to focus on the literature around the idea of nature that directly or indirectly relates to ecological themes and problems discussed in the first chapter. However, earlier work on conceptualisation of nature in Eastern thought in general deals with certain ideas of nature in Classical Philosophy as well as some practices and customs around nature prevalent even today.

As mentioned earlier, two significant ideas frame the background of this summary of works. One is the survey of literature available around conceptualisation of nature in both Western and Eastern traditions of thought and the second is the relevance of such conceptualisations for deriving a possible framework for ecological ethics. What are the conceptualisations of nature we have around us that help us understand what nature is? What is our relationship to our environment?

M. Baindur, *Nature in Indian Philosophy and Cultural Traditions*,
Sophia Studies in Cross-cultural Philosophy of Traditions and Cultures 12,
DOI 10.1007/978-81-322-2358-0_2

How do we move from the conceptualisation of nature that favours exploitation, to one that inspires and encourages conservation? What kind of conceptual ideas or perspectives of nature would create a framework for an ecological ethics that would be free of the various problems and issues discussed in earlier chapter? Philosophical ideas around nature are usually placed within some environmental ethical positions describing a theory in terms of "bio-centrism", "ethnocentrism", "anthropocentrism", etc. These are newer ways of understanding the earlier ideas about nature. My overview from a philosophical perspective ignores these positions (which are sometimes political) and tries to place the conceptualisations within a historical context. Not to do so would exclude the place of Indian thought that does not fall under these divisions due to the unique conceptualisations of nature that are available in these sources. I believe that philosophy, especially Environmental Philosophy, cannot be disembedded from the larger concerns of the society. However, it is also important to have some category to place this history of the "nature idea" in perspective. Therefore, I have chosen to highlight the conceptualisations of nature as "inclusive of human beings" or "exclusive of human beings" (non-human nature), as well as trace the changes in conceptualisations that are seen in various intellectual periods of historical importance.

A possible question that one may ask here is about the time periods of both Eastern and Western thoughts taken up for discussion. Why does the summary of work in Western thought cover a range of conceptualisations from ancient Greek Philosophy right up to the modern philosophers while the predominant discussions of nature in Indian thought clusters around pre-modern concepts?

In this regard, I can only offer the explanation that the historical trajectory of a particular line of Greek thought lead to the gradual development of natural philosophy in the West; in contrast Eastern thought focused on conceptual paradigms of particular traditions that they refer to as principles or *tattva*s. According to Comford (1997), the scientific tendency can be traced through a line of thinkers of the Milesian school and their successors such as Anaxagoras:

> These atomists succeeded in reducing physis to a perfectly clear conceptual model such as science desires, comprised of little impenetrable pieces of homogenous' matter,' with none but spatial properties—tiny geometrical solids, out of which all bodies, whatever shape and size, could be just built up (p. 144).

In Indian thought, though there were schools that had similar ideas, they never gained enough prominence to displace dominant forms of what we know as the Vedantic schools of thought. These include non-dualism, dualism, and special dualism in the later periods of history. Some of these concepts continued to flourish in their embedded forms in the areas of philosophy, language, and other traditions of knowledge such as Āyurveda or Tantra. Also, it is evident that rituals and traditions continued to evolve in practice along with discourses of nature being perpetrated by mythology as well as the narratives of the divine. Some contemporary thinkers such as Aurobindo, J. Krishnamurti, or Tagore did write about nature, but given their background in Western thought, they were already presenting a globalised view, which was also influenced by Western thought.

To prepare a kind of a map that looks at the conceptualisation of nature in Western traditions of thought is to actually trace out the relevant concepts from a conglomeration of disciplines such as geography, history, biology, ecology, and philosophy. Any attempt to provide a complete picture of the concept of nature or a historical description of this idea in the West is too monumental a task and is likely to fall short on many accounts. On the other hand, an engagement with nature from a single perspective—such as the uses of nature in a particular literature genre—could be dismissed as an enumerative project. Lovejoy (1927, p. 444), who has listed the aesthetic uses of the word nature, calls it the "verbal jack of all trades". Therefore, my introduction to the concept of nature deals with those conceptualisations of nature that would make sense in the framework of a philosophy of conservation and action.

At the outset, it is important to remember that concepts are better understood as representations and not definitions. It is true that conceptual resources have their own history of development and have been used or re-fashioned over different time periods and the concept of nature is no exception. On one hand, a concept as distinguished by various traditions of thought and writing can be surveyed; on the other hand, it is also important to be clear about the undistinguished presuppositions that the concept as such is already surrounded by. Though the term "nature", as we use it today is derived from a particular discourse in Western philosophy, there is nothing particularly Western or Eastern about the natural world that we inhabit.

A large part of early Western philosophy was but the philosophy of nature. However, there have been dramatic alterations in the way nature has been perceived. So, as a preamble to my work I think it is important to give a picture of these transformations here. These historical and cultural perceptions of nature have not only influenced scientific thought, but have also been influential in many ways on the development of sciences themselves. Collingwood (1945) in his book *Idea of Nature* traces the historical development of the idea of nature, from the early Greeks to the modern scientific period. A reading of this very detailed book indicates that the concept of nature developed from a long line of philosophy, beginning with the Greek tradition, but we also know now that its meanings and paradigms have changed over time. However, legacy of these early thinkers to Western civilisation cannot be under-estimated. Though the early Greek philosophy subscribed to an idea of nature as a living being, only certain trends in the Greek tradition were adopted into Western Metaphysics and are important to discuss here.

2.2 Cosmological Accounts

Collingwood (1945, p. 29) explains how the Greek conceptualisation of nature in Ionian thought answered the question "what is Nature?" in terms of the query "what kinds of things exist in the natural world?" He further adds that the word nature, which often refers to the collection or aggregate of natural objects in modern sense, was used by the Greek philosophers to refer to a principle, principium, or source. Similar kinds of analysis are undertaken by many traditions of Indian Philosophy to list the possible existents and their relationships. These will be discussed in the later chapters of the book in detail (c.f. Chap. 3).

The word in the roman language, "*natura*" comes from the word "*gnascor*", a root that is the same as "origin" or "birth" (same as the root of the word "pregnant"). However, the Romans, according to Tellenbach and Kimura (1989), used this word to translate the Greek meaning of the word physis. According to Naddaf (2006), the word physis represents the essential nature of a thing—the entirety of its creation, growth, and maintenance: "In sum, phusis [physis] must be understood dynamically as the 'real constitution' offered thing as it is realised from beginning to end with all of its properties". But, Naddaf adds that "physis" is never employed in the sense of a static principle. The original meaning of the word phusis [physis] comprises three types of emphasis—as origin, as processes, and as a result. Psyche was the soul that all the living beings possessed. Homer uses this word in the sense of life. Later thinkers like Aristotle refer to this as "soul" or the "life principle". A point that the philosopher McClure (1933) raises in his paper on the Greek concept of nature is that the "psyche" of nature and man are not the same: "The soul of man is a fragment of the universal fiery energy that appears in the universe at large. Man is the universe in the miniature" (p. 120).

On examining the development of the concept of nature in the Greek period, one notices that each of these early philosophers attempted to answer the question, "what was the essence of the world?" differently. This essence, the undifferentiated stuff for Thales was "water". Collingwood's (1945) description of the Greek concept of nature also attributes the idea of an "ensouled world" to Thales. Anaximenes made out this essence of reality to be air and Heraclitus decided it was fire. The first idea of nature as order seems to come from Heraclitus, where opposites in nature regularly follow each other—such as night follows day and death follows life. Laws of Nature, according to Anaximander, are all nourished by one divine law. This early concept was developed slowly into the sophisticated presupposition "laws of nature" of modern science. McClure (1933, p. 121) suggests that in Heraclitus' cosmology, the idea of a process of tension, a sort of a "give and take" generates the concept of law and order. Carone (2001) explains this "harmony of tension" as something that is inherent in nature for Heraclitus. Marshall (1992, p. 68) writes: "Heraclitus stands at the source of Western metaphysical tradition which stresses process and flux in nature". For Heraclitus, the orderly succession of events was ensured by "reason" or "destiny".

Elsewhere, nature is considered to be moderation (Sophrosune) as opposed to human arrogance (hubris) and nature thus teaches humankind a lesson in humility (Carone 2001). The Greeks greatly appreciated the beauty and order in nature and any disturbance of the natural order was also "hubris" and attracted punishment from the gods.

For Ionians, there was a distinct problem of how to explain the differences in natural things. If everything was "stuff" and homogenous, how would different things have different properties? Also, it was not clear how one could distinguish matter from the space it occupied. The resolution of these questions was the philosophy of Pythagoras and his school which is always identified with him. Collingwood (1945), writing about Pythagoras, states that this ancient Greek philosopher was trained as a mathematician; therefore, he attempted to find a connection between the problems of cosmology and the achievements of geometry. Pythagoras suggested that qualitative differences in nature were based on differences of geometric structure. It is not therefore important to ask what primitive matter is like or ascribe to it any character differing from that of space itself. All we must describe is the power of being shaped geometrically. The nature of things by virtue of which they severally and collectively are what they are, is geometrical structure or form. Collingwood (1945) further suggests that the Pythagorean project continues into present day science which correlates properties with structure. So according to Pythagoras, it does not matter what the world is made up of; what we have to strictly look at is the patterning and changes of matter which provide the explanation of properties.

Carone (2001) points out that in later Greek thought of Socrates, the order of the universe was important. The idea that this order (both natural and social) was maintained by friendship and moderation seems to be repeated as a constant theme in his speeches. Plato, considered the disciple of Socrates, is said to adhere to what is popularly called "metaphysical dualism". Carone (2001) also suggests that, in order to deal with the flux of the phenomenal world, Plato invokes the concept of form, the stability of which allows us to attain stable knowledge. While the perceptible world is in a state of flux, the ideas or forms are unchanging. Collingwood (1945) explains the relation between the forms and the perceptible world for Plato:

> The world, the aggregate of natural things was throughout its fabric a complete of matter and form. Form was wholly immanent in the world. Form the intelligible had its being only as that which re-entered intelligible the world in which it was immanent (p. 72).

Timaeus contains some of the main ideas of Plato's cosmology. Here, one finds that the emphasis has shifted, from the idea of matter to the idea of form. Plato's cosmology is like a logical succession to Pythagoras's concepts and the notion of mathematical form is extended to include all other forms. "Matter" in the *Timaeus* is simply that which is capable of assuming geometrical form, and the form which it can receive is independent of any material embodiment and constitutes itself and apart from the matter in an intelligible world. The world of nature is a material organism or animal, alive everywhere with spontaneous movement. The intelligible world is called the immaterial organism, alive, because forms are related dynamically in dialectical connections. The same world is not alive with movement, because movement implies space and time and the world of forms has no space and time. If space and time are not present in the world of form, where do they arise in the world of nature? Since the world of nature is a copy of the world of form, one would expect that every feature have a corresponding ideal made in that world (Collingwood 1945).

According to Collingwood (1945), the *Timaeus* suggests that space corresponds to no feature of the intelligible world.

> Space is simply that of which the copy is made, like it is the sculptor's clay or draughtsman's paper. In the intelligible world, everything realises its entire nature simultaneously—for example, all properties of a triangle are present in the triangle at any given movement, so it does not need a lapse of time to realise then one after another. In the perceptible world the total nature of a thing is never realised all at once (Collingwood 1945, p. 80).

It is in Plato's philosophy that we begin to find early ideas of the one cause which Aristotle engages with as the prime mover. *Timaeus's* answer to the question about the need for god as a creator is that the world of nature is a becoming process and that all things becoming must have a cause. Why should god have created this world? The reason that Plato gives is that god is good and the nature of goodness is to overflow outside itself and reproduce itself.

Citing evidence from Plato's works that are available to us, Carone (2001, p. 70) also discusses Plato's idea that humans who have not lived up to the highest rationality may be reborn as animals. It is also interesting to note that for Plato, plants too are like humans and that they experience pain and pleasure. This indicates the idea of a shared psyche. It is easy to naively conclude that "physis" and "psyche" are placed in opposition in Greek thought, but one must be careful to remember that psyche is inherent in physis and not necessarily opposed to it.

Aristotle is often considered Plato's successor. His view is that nature is a source or cause of things being moved and of being at rest, within that to which it belongs primarily. In his work *Physis II*, Aristotle (Trans. 1963) begins by talking about the various ways in which nature can be defined:

> Of the things that exist, some exist by nature, others through other causes. Those that exist by nature include animals and their parts, plants, and simple bodies like the earth, fire, air and water—for of these and such like things we do say that they exist by nature. All these obviously differ from things that have not come together by nature; for each of them has in itself a source of movement and rest (p. 209).

Natural things, once their generation begins, "grow by means of themselves" (p. 210). This is the most fundamental difference between art and nature for Aristotle; nature is a source of change within the thing itself. Two concepts were important in Aristotle's view of nature: purpose or teleology and the order or hierarchy of beings. Aristotle's concept of "*entelechy*" demonstrated that nature had a tendency to move towards definite ends. Nature had a purpose. Inherently, all beings occupied a place within an order of hierarchy. Coupled with the idea of cornucopia or abundance of beings in nature, Greek philosophers could construct a scale of beings from the highest to the lowest. The beings lower in order fulfilled their purpose by satisfying those higher up in the hierarchy. The value of a being was fixed by where it was placed in this order.

2.3 Beginnings of a Science of Nature

For a long time, the Aristotelian concept of nature held sway in sciences and philosophy. In the sixteenth and seventeenth centuries, there was a movement away from Aristotelian physics in what Collingwood (1945, p. 94) perceives as a major cosmological revolution. There was a shift away from teleology and purpose to the enquiry into the process structure as well as nature of the world:

> The naturalistic philosophy of the Renaissance regarded nature as something divine and self-creative; the active and passive sides of this one self-creative being was differentiated by distinguishing 'natura naturata', or the complex of natural changes and processes, from natura naturans, or the immanent force which animates and directs them (1945, p. 95).

This important idea in medieval times was the shift from the telos of the world, towards a telos that made man central in every sense. Brutt (1924) also writes of this shift:

> For the dominant trend in medieval thought, man occupied a more significant and determinative place in the universe than the realm of physical nature...the whole world of nature was believed to be teleologically subordinate to him and his eternal destiny (p. 4).

Brutt (1924, p. 5) also points out clearly as to how two movements led to this conviction: Greek philosophy (post Aristotle) and Judeo-Christian theology. He connects this underlying world view of man's all-important and influential role in the universe to the development of medieval physics. He posits that not only was nature subservient to man under this conceptualisation, but the world of nature was also considered to be "immediately present and fully intelligible to his mind". Post Aristotle, Greek Philosophy in the early phase of Renaissance still believed that the world of nature—*natura naturata*—was like a living organism. Nature could feel pain and pleasure, or experience love and hate. With the coming of the new astronomy of Kepler and Copernicus, the idea of nature as an organism was replaced by the idea of nature as a machine. It has also been argued that the idea of nature as "designed" became prominent as the orders of its workings were tabulated. To a certain extent, the design theory of nature actually helped to divinise nature, as there was a stronger argument for a divine mechanic who put nature's clockwork together. The impact of this idea was that "nature" was no longer a copy of some other original perfect world, and this was evident in the way it was studied and recorded. Collingwood (1945) points out the shift in the way nature was perceived:

> From an early date in the history of the movement it led people to think of nature as self-creative and in that sense divine, and therefore induced them to look at natural phenomena with a respectful, attentive, and observant eye; that is to say, it led to a habit of detailed and accurate observation, based on the postulate that everything in nature, however minute and apparently accidental, is permeated by rationality and therefore significant and valuable. The Aristotelian tradition, regarding nature as a material imitation of a transcendent immaterial model, implied that some things in nature were accidental (p. 95).

We can say that conceptualisation of nature in the medieval period based itself firmly on two intersecting ideas that came together. One was the concept of the designed universe with a divine creator and the other was the purpose of this creation, which was to be used by man quite firmly. Also embedded in the idea of man's geo-centrically located importance was the theological movement. The prime mover of Aristotle and the personal father in heaven became one, and this god favoured man. The entire realm of nature was created by god for man to know and enjoy. Leiss (1994, p. 31) states that: "science conceived as the winning of mastery over nature seemed to be the natural fulfilment of the Biblical promise that man should be lord of the earth". The famous passage in the Bible reads:

> And God said; Let us make man in our image, after our likeness: and let them have dominion over the fish of the sea, and over the fowl of the air, and over the cattle, and over all the earth, and over every creeping thing that creepeth upon the earth (Citation).
> So God created man in his own image, in the image of God created he him; male and female created he them. And God blessed them, and God said unto them, be fruitful, and multiply, and replenish the earth, and subdue it; and have dominion over the fish of the sea, and over the fowl of the air, and over every living thing that moveth upon the earth (Genesis 1:26–28)

We can trace an important development in the Christian Theology that created a dichotomy in the concept of nature in the popular imaginations of that period. While all nature was divine and sacred, there also seems to be a parallel idea of a wild and savage nature that resisted domination. The spirits and forest creatures were not gentle, nor were they "usable" by human beings. The sacred creation of God became less sacred in the wild as it resisted man's domination. Leiss (1994) traces this to the myth of man's dismissal from Eden. With the fall of man from Eden the most sacred garden, nature too was subjected to a fall from the sacred place it occupied. This lead to the conviction that the more nature was tamed and brought under human control, the more earth would be like paradise. The idea of imitating this lost paradise also lead to the development of gardens, cultivated landscapes, and domestication of nature both in Islamic and Christian cultures at this time (Leiss 1994, p. 31). Leiss writes about this:

> The decisive question for Christian theological commentary on this point was how the Fall affected man's dominion on earth. The existence of wild animals was regarded as evidence that there had been a partial loss of authority on account of sin, for it was assumed that in the Garden of Eden all animals obeyed man's bidding. The domestication or destruction of the wild animals would be a sign that the earthly paradise had been restored. The legends recounting the deeds of the early saints who retired into the wilderness all speak of their accomplishments in taming beasts as proof that they were reasserting the rightful human sovereignty enjoyed prior to the Fall (p. 31).

On one hand, nature would be subject to use by man, sanctioned by the divine. On the other hand, investigating nature would fundamentally imply investigating the secrets of the divine plan and order of things in the universe. Leiss (1994) further points out:

> Nature has a double aspect. In its immediate presence, as the source of satisfaction for vital human needs, it necessarily arouses utilitarian modes of behaviour (which may differ

> widely in structure and detail); reflectively, however, nature appears as the visible testimony of God's providence and thus must be regarded from the perspective of its value as an aid in understanding the divine scheme (p. 34).

During the Renaissance, there were steady successions of remarkable technological innovations. Many contemporary scholars have claimed that the Renaissance is the primary modern source for the idea of mastery over nature. Alchemy, magic, and other investigations into the nature of matter played a role in furthering the cause of man's mastery of nature. The idea that these arts were against god and one was interfering with nature if one engaged in science and technology was perhaps averted by the timely work of Bacon. He was universally praised as "the secretary of nature". Bacon (1620, quoted in Leiss 1994) provided the world of scientific method with legitimisation, whereby the idea of mastery over nature became widely acceptable and the cultural impact of religion gradually diminished. The concept of mastery over nature has been regarded as an outstanding contribution of Bacon's world view for a long time. Through explaining the concept of "The Fall from Eden", Bacon establishes his related contention that trying to know nature would not only give us complete understanding of god's plan but would reverse the effect of the fall from Eden. He writes of the restoration at two levels:

> For man by the fall fell at the same time from his state of innocency and from his dominion over creation. Both of these losses however can even in this life be in some part repaired; the former by religion and faith, the latter by arts and sciences (1620, quoted in Leiss 1994, p. 31).

The distinction that man's innocence could be reversed by faith and religion and that the dominion over nature could be restored through arts and science, particularly had the effect of secularising the field of human endeavour that was based on "understanding nature". Not only was the development of this thought limited to the terrestrial realms, but even as a quest for understanding nature, the discovery of the laws of nature brought under human understanding even the celestial bodies which had eluded reason. Merchant (2004) points out: "Recovering the lost Eden became Western culture's major project during the scientific revolution of seventeenth century. Reason and experiment were keys to reinventing Eden on earth" (p. 65).

Post-Renaissance, the scientific revolution is considered to be the source of the dominant world view that saw nature as dead and mechanistic. This view paved a way for not just domination of nature, but through the development of technology, it brought in the discourse of "harnessing nature". We have to consider two areas differently from each other: the development of the natural sciences and the development of technology. Both depended on the transformation of the concept of nature. It seems that the realm of metaphysics was separated from physis, while technology succeeded in maximising extraction of materials from nature. It was not that nature was not "used" earlier, but that somehow the moral values that it was accorded with disappeared. Marshall (1992) writes of this shift:

> It marked a fundamental shift in our relationship with the natural world, which was no longer considered a divine dwelling for humanity, but an object to be used. By insisting on a rigid split between the observer and the observed, it further alienated man from nature.

> No longer fearful of disturbing the vanished gods, he could exploit the machine of nature to the hilt. Released from earlier moral and religious curbs, he felt free to maximise his power in the untrammelled pursuit of his own ends. In a double process, it not only desanctified nature, but also gave man enormous power over it (p. 168).

Marshall (1992) also points out that the attitude related to the devaluing of nature is instantiated by the behaviour of the people during that age towards animals. There was also the development of the positivistic philosophy of economics, which recommended human material wealth as the ultimate aim of human happiness.

2.4 Heliocentric Revolutions

The revolution in science referred to as the Copernican revolution, changed not only the way the Western civilisations perceived the world, but also changed the way human beings looked at themselves The new heliocentric astronomy of Copernicus and Kepler had profound influence on the concept of nature. Collingwood (1945) argues that the popular perception that the work of Kepler reduced the importance of the earth and the human being in the scheme of things is wrong. He writes:

> This idea, so far from diminishing the scope of man's powers, vastly enlarged it; for it taught him that scientific laws established by him on earth would hold good throughout the starry heavens. It was directly owing to Copernicus's denial of geocentric astronomy that Newton could imagine the force which kept the moon in its orbit to be the same that drew his apple to the ground (pp. 97–98).

Burtt (1924, pp. 38–39) claims that certain presuppositions about nature lead to the acceptance of Kepler and Copernicus' new astronomy. One was that nature was governed by the principle of simplicity. Following Galileo, the teleology of the world and creation, in other words the "why" that lead to finding primary cause, was replaced by the investigation into the processes and laws or "the how" of nature. Kepler's explicit analogy between nature and a machine is clear in this section of a letter written by him:

> I am much occupied with the investigation of the physical causes. My aim in this is to show that the celestial machine is to be likened not to a divine organism but rather to a clockwork … insofar as nearly all the manifold movements are carried out by means of a single, quite simple magnetic force, as in the case of a clockwork all motions [are caused] by a simple weight. Moreover I show how this physical conception is to be presented through calculation and geometry (Kepler's letter to Hewart Von Hohenburg, quoted in Marshall 1992, pp. 171–172).

After Kepler, the next idea that changed the concept of nature was given by Galileo, who suggested the idea of inertia in physics and did away with the need for an agent like God to "keep moving things". These discoveries had a profound impact on the concept of God. He became a divine mechanic, no longer perceived as the powerful 'Prime Mover.' The long line of thinkers thus added to the concept that nature was knowable, measurable and interpretable in mathematical terms. The genius among these thinkers was Newton, claims Burtt (1924, p. 202). Speaking about Newton's achievement, he writes: "For him to invent the needed tool and by its aid to reduce the major phenomena of the whole universe of matter to a single mathematical law…." This victory of understanding is aptly summed up in Alexander Pope's poem meant as an Epitaph intended for Sir Isaac Newton.

Nature and nature's laws lay hid in night;
God said, "Let Newton be!" and all was light.

Burtt (1924, p. 17) makes an observation that this was the time when natural philosophy and natural science split and "… it is largely due to Newton that a real distinction came to be made between the two". On the other hand, at the same time, philosophical movements that countered this mechanistic view of nature were also present. The movements of romanticism and those of the anti-Enlightenment did make their presence felt at various periods in history.

The legacies of the two great cultural movements which have dominated Western thinking since the eighteenth century are ambiguous and their full repercussions are still being worked out, claims Marshall (1992). These two movements in Western thought were Enlightenment and Romanticism.

Post-Renaissance, the question about the extent of enterprise of human reasoning lead to the development of a period of philosophy known as the enlightenment. Hibben (1910) writes about the significance of this historical development in Philosophy.

> The significant movement of thought known as the Enlightenment, or *Aufklärung*, falls in the main within the period of the eighteenth century. However, it is seldom that the turn of a century happened to coincide exactly with the beginning or the end of a great epoch, political, religious or philosophical. The period in philosophy which is referred to in a general way as the eighteenth century begins virtually in the year 1690 with the publication of Locke's famous work, An Essay Concerning Human Understanding, and is brought to its close in the year 1781 with the appearance of Kant's The Critique of Pure Reason. They are the natural boundaries of this "philosophical century (p. 3).

During the Enlightenment, there was a spread of liberating thought that freed humankind from superstitions and encouraged free thinking. The challenge of authority—of church, monarchy, or forms of tradition—resulted in the superior place for reason in the world. During Enlightenment, the idea of progress became stronger. The belief that nature would progress along with mankind was often presupposed due to the elevated status of the human being. Faith that was Christian gave way to a humanist faith. Marshall (1992, p. 294) writes: "The growing mastery over nature through technology and science made man more arrogant than ever". The apriori, analytical reasoning of the Enlightenment replaced all other kinds of thinking and though it was reason that was to prevail, only instrumental reason prevailed (p. 295). The critique of this period is based on the fact that humanist-centred perspective, particularly a man-centred perspective tended to ignore the rest of the natural world including and other kinds of thinking, intuition, emotions, or feeling.

2.5 Romanticism

Temporally, the idea of evolution replaced the idea of human history. Earlier classical civilisations such as the Greek and the Romans perceived their relationship with nature conditioned by their prevailing belief that human history is cyclical or a meaningless flux of events. This was replaced by the Christian world view of sin, redemption, and a move towards salvation. Very subtly, the Christian concept of forward movement integrated into the Enlightenment idea of "progress".

In response to the Enlightenment period, a tradition which denied the objectivity of science and the split between the observer and the observed developed. Popularly referred to as "the Romantic Movement", this revolt against the Enlightenment and the Positivist sciences sought to break away from the idea of a quantitative vision of nature, towards a more qualitative vision. Thinkers of this stream of thought looked for ways of knowing that would avoid objectifying nature completely. The concept of an organic nature replaced the idea of a clockwork or machine. The idea of nature in poetry and writing, which persists to this day in wilderness debates, saw their beginnings during the Romantic period. The presupposition in such literary works was that nature was pure and it was needed to cleanse and purify humankind of the ills of an obsolete and corrupt society—one that was artificial and mechanical. Nature's innocence and goodness creates a powerful value-based separation of places—such as the countryside or wilderness that was ideal and the built up towns and cities that were polluting. Marshall (1992, p. 285) quotes Whitehead on the Romantics: "As the philosopher A.N. Whitehead argued, one of the most important lessons for modern science was the Romantics' protest on behalf of the organic view of nature, and … against the exclusion of value from matter of fact".

Schelling (1775–1854), generally regarded as the principal philosopher of Romanticism, posits an ecological world view in which the description of nature is not restricted to scientific explanation. Schelling responds to the Kantian project in his work *Naturphilosophie*. Kantian thought and idealism posits that a "thing in itself" can never be known and all that is known is our representation of it. Though in this stream there is a denial of the reality of a nature apart from our knowledge of it, there is no ontological denial of the existence of a nature outside us. "The Kantian division between 'appearing nature' and 'nature in itself' is seen as resulting from the fact that the nature theorised in cognitive judgments is objectified in opposition to the knowing subject" (Bowie 2001). According to Schelling, this objectification of nature fails to account for the dynamic forces in nature, including our own selves. He argues that our concern should not be about the nature that is actual and unavailable to us, but about what actualises nature for us or how we "construct nature". Nature then exists for us as "a possibility of nature". Experiments and other empirical sciences only provide us with a surface concept of nature: Every experiment contains an implicit apriori judgment of nature, making it a "production of nature".

For Schelling, a new science—that he refers to as 'speculative physics'—would engage with nature without these aproiri judgments. He bases his theory of nature on the fact that the essence of matter consists of opposing forces of attraction and repulsion. The *Naturphilosophie* includes humankind within nature, as part of an interrelated whole: "This philosophy must accept, therefore that there is a hierarchy of life in nature. Even in mere organised matter, there is life, but a life of a more restricted kind" (Bowie 2001).

2.6 Evolution

Historical and mythological explanations of change and the origins of life forms were replaced by the development of the early biological sciences and these new ways of thinking had a major impact on the perception of nature. The travellers, who first embarked on journeys around the world, brought back tales and creatures from far off exotic lands. On one hand, we had the coming of the "wild" and the "native" worlds discovered by the Western anthropologists; on the other hand, with the high number of life forms that were discovered, taxonomy ruled the largely descriptive project of natural history until the work of Darwin.

The concept of evolution of life forms is often considered to be the "Copernican" revolution in the life sciences. The work of Darwin and others had the effect of completely changing the fundamental understanding of people with regard to life forms. Firstly, the special position of man among all other beings on earth was displaced. The naturalistic explanation linked the human beings to the rest of the natural world as human beings took their rightful place in the evolutionary tree of life. Darwin's work is succinctly summarised by Marshall (1992) thus:

> Darwin's theory of evolution through natural selection, the most unifying of all biological theories, is based on three observations and two deductions. The first observation is that organisms tend to increase at a geometrical rate; the second, that the populations of different species are more or less static. From this, he deduces that vast numbers of organisms die before they can reproduce. The third observation is that there are inherited variations between the same members of a species. The final deduction is that in the struggle for existence, those variations which make the organism best adapted to its environment will give it a better chance to survive and reproduce. The result will be the gradual evolution of different species and the formation of new species (p. 323).

Darwin's works show that he had a very holistic view of nature and was moved by its complexity and beauty. Darwin's reference to nature in his work and the romantic tendency to see nature in terms of a nurturing mother has resulted with reference to the idea that is referred to by a few as the "hidden Goddess of Darwinism". Darwin was also very clear about the interrelationships between various parts of the natural world. He recognised the concept of places for organisms to occupy (which we now call 'niches'). Stressing the importance of the concept of evolution, Marshall (1992) states that the work of Darwin changed the absolute position of the human and the scale of ethics, by implying that man's position in the world of nature was only a matter of degree. He writes:

> The most important legacy of the Darwinian revolution is that it has undermined once and for all the attempt to defend the notion of human uniqueness on the grounds of intelligence. Humans are social animals who have developed reason, speech and a moral sense more than their fellow animals (p. 331).

Both Kant and Herder also influenced the idea of nature through their philosophical thought. Tracing the philosophical idea of cosmic evolution, Marshall (1992) writes of the influence of Kant and Herder: "Immanuel Kant too in his *General History of Nature* and *Theory of the Heavens* (1755) interpreted the whole universe as a product of historical development and offered the first systematic evolutionary account of cosmic history". His pupil Gottfried Herder (1744–1803) put forward a theory of organic development of cultures. Herder visualised nature as a great chain of being which placed all beings in a hierarchy, but at the same time, various beings were not separated from each other, as they were all part of an organic whole, interrelated, and interdependent a dynamic and evolving web of life. From observation of nature and its cosmic linear movement from chaos to order, one can deduce the evolution in history.

Marshall (1992) also points out that this concept of nature as consisting of a great chain of beings was positive for ecological thought. He claims that "doctrine of the Chain of Being continued to see diversity as a form of excellence" and that this chain related the human being to all of other beings in the universe by a process of evolution over time: "With Herder, the Great Chain of Being is thus temporalised: the passage from 'lower' to higher marks a succession from 'earlier' to 'later'".

Referring to the doctrine of the Chain of Being as inherently animistic and a natural ally of Pantheism—which presented the world as the overflowing manifestation of the One—Marshall suggests that Herder's philosophy paved the way for an evolutionary and ecological appreciation of nature and Romanticism. It is important to note that although it seems that Herder would almost anticipate Darwin, he in fact did not conclude that creatures have evolved from earlier ones by descent. Herder adhered to the Judaeo-Christian view of man's privileged position as lord of the creation. The earth was made for man, who was the privileged lord of all creatures on earth.

It is important to note that the development of natural science on one hand and the study of human problems and issues by historians on the other contributed to a conception of nature based on history. Nature became progressive. The idea of progress was simultaneously adopted for the human being and nature. "…civilization has moved, is moving and will move in a desirable direction" (Bury, cited Lafrenniere 1985). With the development of the field of biology, the growth of natural history and concepts from social sciences were borrowed and projected on to natural sciences and vice versa.

With the wide spread acceptance of Darwinian notions of evolution, nature ceased to be stable and constant. "Nature was no longer perceived as a machine"

(Collingwood 1945); instead nature that was organic or biological was perceived as something that could replicate and evolve. On the other hand, the utilitarian mode of perception took over when nature was considered the source of raw material for human use. The concept of evolution thus marks the next great leap in the concept of nature.

2.7 Modern Cosmology

On the metaphysical level, the impact of the theory of evolution was the discussion of emergent properties. Two thinkers that Collingwood (1945) chooses to describe who worked on a modern cosmology are Alexander and Whitehead. Alexander's idea of nature was based on the view derived from the concept of "emergent evolution": "This world, as it exists in its ceaseless changes, appears to him as a single cosmic process in which there emerge, as it goes on, higher orders of being" (Collingwood, 1945, p. 158).

This view proposes that the evolved state of anything is not a mere product of an improvisation or modification of the original, but the new patterns emerging in the new state represent a new order of being, whose properties cannot be explained by being reduced to earlier properties of the evolutes. This idea was proposed for the concept of life by thinkers such as Morgan, who saw life as an emergent property not reducible to properties of mind or that of matter. Life and mind thus cannot be the subject of just Biology reducible to physics. Alexander extends the idea of the new patterns giving rise to new properties to the whole of creation. Collingwood (1945) writes about Alexander's view on living organisms using the concept of space-time in a unique way:

> Living organisms in their turn are patterns whose elements are bits of matter. In themselves these bits of matter are inorganic; it is only the whole pattern which they compose that is alive, and its life is the time-aspect or rhythmic process of its material parts. Thus life is the time-aspect of the organism, its space-aspect being inorganic matter; in other words, life is a peculiar kind of activity or process belonging to a body composed of parts which taken in themselves enjoy an activity of the next lower order (p. 160).

For Alexander, different orders of being within matter exist, "the higher being elaborated forms of the lower, and different orders of mind" (ibid.). The formation of higher order of categories with more complexities causes this infinite evolutionary process. Yet, each of this higher order complex emergent forms is take on a form of a new simple totality that may form the basis for the next stage of evolution and so on. Collingwood (1945) suggests that this cosmic process envisioned by Alexander rests on a presupposition of categories derived from the definition of space-time:

> Space-time is the source of the categories, but they do not apply to space-time; they belong only to what exists, and what exists is not space-time itself but only the empirical things in it; but these things possess categorical characteristics for one reason and one only—namely, that they exist in space-time (p. 162)

Collingwood (1945) suggests that Whitehead's concept of nature is similar to the one proposed by Alexander. Whitehead's early training was that of a mathematician and physicist: "Nature for him consists of moving patterns whose movement is essential to their being; and these are analyzed into what he calls events or occasions, which correspond with Alexander's point-instants" (p. 165).

Similar to Alexander, Whitehead also believed that the structure of a being gave it its properties and that a breakdown of structure and further analysis of it would reveal components, but would destroy this structure that gave an object its unique properties. Thus, Whitehead posits that everything that exists is like a living organism because its essence depends not only on its individual parts, but on the particular combination or pattern in which they are organised. Explaining Whitehead's ideas on the concept of nature in detail, Collingwood (1945) suggests that the processes and the organic form of nature as an organism are separate:

> The activities of the organism are not external accidents; they are united into a single complex activity, which is the organism itself. Substance and activity are not two, but one. … The process of nature is not a merely cyclic or rhythmical change, it is a creative advance; the organism is undergoing or pursuing a process of evolution in which it is constantly taking new forms and producing new forms, in every part of itself" (p. 167).

The cosmic process of nature has two parts: one is "extensiveness" in space-time and the other is "aim", which is a process explained by teleology. This is given by the idea that a process is always oriented towards a particular goal. Alexander, being an empiricist, believed that the emergent qualities of the new patterns would solely abide in the patterns; Whitehead however believed that these qualities belonged to an eternal platonic world (Collingwood 1945).

2.8 Phenomenological Traditions

Phenomenology concerns itself with the experience of human beings as they find and construct meanings in the world of everyday encounters with it. Edmund Husserl is credited with creating a way out of the paradigm of purely naturalistic view of things. According to David Wood (2005), phenomenology is a resistance movement against treating phenomenon as reducible to causal laws.

Historically, the idea of nature seems to have taken two trajectories of conceptualisation. One concept is that it is the object of study for the natural sciences. The other is that of nature being the "natural world" that we live in and the one that we experience through our senses. Husserl considered the everyday world of experience as the "life-world" (*Lebenswelt*). This is the world of familiar encounter with the environment, not in terms of any epistemological sense. This world is founded on everyday human understanding and is pre-scientific, as no operations of abstraction take place.

Husserl posits that when we do science, there is partial departure from the life world into a different world of natural scientific objects that are unchanging and subject themselves to mathematical treatment. Nature is idealised, or in Husserl's own phrase, the "mathematisation of nature" takes place. The world of nature that is one of the experiences seems distinct from this abstracted world of objects that are mathematical. Liess (1994) summarises Husserl's brief references to nature in *The Crisis of European Sciences* (1936). He explains the bifurcation seen by Husserl in the apparent distancing of nature as an object of scientific study and nature as experienced by human beings every day:

> Corresponding to the two spheres of human activity in modern life are two worlds of nature: intuited nature (lebensweltliche Natur) and scientific nature (wissenschaftliche Natur), the experienced nature of everyday life and the abstract-universal, mathematised nature of the physical sciences (Liess 1994, p. 135).

The nature of the modern scientific enterprise, according to Husserl, does not possess a being that is subsequently recognised by man; rather, it receives this being by entering into the historical world of man and by being subjected to the experiments conducted by man. Only to the extent that nature can be subject to this kind of operative manipulation can it be said to be "nature" in the sense of being the object of natural science, and only on this basis can natural science become an efficient tool for the technological domination of the world (Husserl 1936, quoted in Liess 1994).

Therefore "nature" per se is not the thematic object of the activity of science, more precisely, the natural sciences. Each perspective on nature finds a different relationship of human interest. While the human relationship to the natural world is often fraught with conflict and struggle, the final benefits of nature being used have resulted in real effects on the society. The gains begotten from the control of nature seems to not reach everyone, leading to power struggles within human beings:

> Along with his growing, more and more perfect cognitive power over the universe, man also gains an ever more perfect mastery over his practical surrounding world, one which expands in an unending progression. This also involves a mastery over mankind as belonging to the real surrounding world, i.e., mastery over himself and his fellow man, an ever greater power over his fate … (Husserl 1936, quoted in Liess 1994).

Husserl's search for a rational foundation for the interaction of the life–world and the scientific world is based on his contention that the natural sciences "hide" the connection between the two natures.

> From the viewpoint of science the nature given in sense perception masks the underlying uniform structure of matter, and modern science's mastery consists in penetrating this disguise and identifying the characteristics of that structure. Considered from the opposite angle—from the viewpoint of life in the familiar world—the mastery of science is manifested in its ability to cast a "veil of ideas" (Ideenkleid) over the nature experienced in everyday existence, that is, to treat the phenomena of nature as if they were purely mathematical-geometrical objects (Husserl 1936, quoted in Liess 1994).

Husserl's work is significant in the conceptualisation of nature, as it addresses the beginning of multiple visions of nature, clarifying the phenomenal from the studied nature in course of Western thought.

The work of Merleau-Ponty, another important philosopher in phenomenology, is founded on the idea of perception. Wood (2005) describes this stream of philosophy thus:

> A phenomenology of perception insists that it is only as spatially and temporally embodied beings that seeing takes place at all. Seeing is made possible by there being discreet bodies, including ourselves that occupying distinct places at particular times, bodies endowed with a mobility that reflects their needs and desires, and bodies whose very discreteness belies a deep interdependence. These are not just natural facts about the word but fundamental dimensions, dimensions that structure of the very possibility of there being facts at all (p. 311).

Through the study of Phenomenology, a number of environmental philosophers have worked on conceptual issues of subjectivity and the experience of nature. Attempts have been made to reverse or at least conceptually oppose the earlier views of naturalism, as well as the conceptualisation of "nature as difference" and also the idea of "nature as origin". Vogel (2005) argues that environmental thinkers working in the post-structuralist traditions tend to think of nature as an idea origin or difference. According to him, these views segregate humans from nature one predates nature's existence from human beings, the other separates nature as non-human. For him, understanding nature as a social construct would actually solve the ontological debates around nature. This makes nature connected to practice. For Vogel (2005), nature is the name that we give to the very concreteness of practice.

The multiple visions of nature in the current age do not lend themselves to a very clear historical discussion. The development of different disciplines around the study of nature has created a profusion of themes that are now broadly referred to as Environmental Humanities. There are two broad themes within this field. One is the continuing study of nature as an object, which led to the development of the various disciplines of environmental studies; and the second is the increased awareness of human relationships with nature in literature and culture: environmental philosophy, ethics, and the fields of nature writing and ecocriticism.

Ecological nature is one of the concepts created by the study of nature as an object. Ecology is commonly known as the study of interactions between organisms and their environment, which includes other organisms. Ernst Heinrich Haeckel (1834–1919) was the first to use the word *oecologie*, in his *Generelle Morphologie der Organismen* (1866), though he was not the first to formulate the principles of ecology. He defined ecology as "The economy of nature". Briefly, we can say that the pattern of distribution, abundance dynamics of organisms, and their interactions are studied in this discipline, at various scales of spatio-temporal resolutions. The development of this discipline had major implications for the conceptualisation of nature.

Ecocriticism, a recent discipline, studies the relationship between nature and literature. The subject is a form of literary criticism that approaches texts through an environment-centred perspective. The fundamental premise of ecocriticism is that human beings are deeply influenced by their environment that consists of both natural and cultural aspects. While earlier literature studies have focused on the relations between human beings and their relationship with each other and society, ecocriticism or literary ecology as it is sometimes called examines human relations with nature as represented in literature. The term "ecocriticism" was coined by William Rueckert in his 1978 essay, "Literature and Ecology: An Experiment in Ecocriticism". He suggests that experimental ecocriticism must address how literature and ecology could develop a symbiotic relationship. He writes: "how can we move from the community of literature to the larger biospheric community which ecology tells us (correctly, I think) we belong to even as we are destroying it?"(p. 121)

Howarth (1996, p. 69) writes that "Ecology is a science strongly connected to a history of verbal expression". Most ecological voices in history have used writing to convey the ideas about nature and the state of the environment. Both naturalists and poets were the creators of a large genre of literature called "nature writing" and they have contributed richly to the themes and debates in ecology. Some of these early writers included Aldo Leopold (*A Sand County Almanac* in 1949), Henry David Thoreau (*Walden* in 1960), and George Perkin Marsh (*Man in Nature*, in 1964).

The first phase of this discipline mirrors feminist criticism in American literature, suggests Glotfelty (1996). In her introduction to this discipline, she writes about how one goes about doing ecocriticism:

> Analogous efforts in ecocriticism study how nature is represented in literature. Again consciousness raising results when stereotypes are identified—Eden, Arcadia, Virgin land, miasmal swamp, savage wilderness—and when absenses are noticed. Where *is* the natural world in this text? (p. xxiii)

She also adds that other topics related to nature and culture such as geographical features, rivers, mountains, deserts, animals, and the body are also identified. Further developments of this field, particularly in American literature has led to development of the genre of nature writing in English. While there was a revival of nature-based writing of both fiction and non-fiction on one hand, there was a renewed interest in mainstream writers whose work manifested ecological awareness. According to Howarth (1996, p. 80) Ecocriticism uses "*deixis*", or the ability to point, to analyse language. He explains: "More developed in Asian rather than European language (liu) deixes locates entities in space time and social context. Through deixes meaning develops from what is said or relative to physical space: I–you, here–there, this–that" (p. 81).

He further adds that ecocriticism seeks to examine how the metaphors of nature are used and abused (p. 81). For studying nature in literature produced in India, one cannot use the taxonomies of nature writing based on the traditions of ecocriticism from the West. Non-fiction literatures—such as natural history, descriptions of walking and rambling in nature and wilderness accounts—all are particular to certain genres of English literature that were deeply influential in the study of nature writing. Instead, in these other literatures, we have writing that is embedded in nature and location. Place-based writing—sacred landscape accounts, pilgrimages, health geographies, travels through the forest, oral histories, and philosophical texts—all of these contain within them descriptions and representations of the natural world in some form or the other. For instance, the topographical classification of landscape is found within a traditional system of medicine, *Āyurveda*. The Jaina texts describe an almost mythical geography of hills and continents and rivers. Kalidasa's poem *Meghadūtam*, a fictional rendering of a cloud's route to the Himalayas, has references to real forests and rivers of central India.

One must add that among the theories of literary criticism is also the category of *tiṇai*, or a reference to particular ecotypes, which in its current form, is only applicable to a particular genre of poetry in the Tamil language. More work is required in all these areas, particularly from the growing numbers of environmental philosophers in India.

Love (1996, p. 237) argues that there has been widespread rejection of writing about nature: "The literature in which nature plays a significant role is by definition irrelevant and inconsequential". He suggests that the recognition of the ecological perspective is not unique to Western American literature or certain regions. He clarifies that "[e]cological issues are both regional and global. They transcend political boundaries. What is required is more interdisciplinary scholarship and more interregional scholarship on common issues" (p. 237). He also emphasises that literature should direct its attention towards recognising "… the primacy of nature, and the necessity for a new ethic and aesthetic embracing the human and the natural…" (p. 237)

Allen (1996, p. 241) suggests that there have been issues about applying principles of ecocriticism to literature that is non-Western. She points out that there is a tendency to club together all these literatures under the theme of "folklore" and call them "primitive or pagan" (p. 241). Using the example of American Indian literature, she suggests that it is important to study them from the perspective of the people who produce these literatures. Understanding the culture would bring to the forefront the complexity and richness of meaning in these narratives of nature. Particular to these cultures is the unity of the human, non-human, and the supernatural world in ceremonial literature:

> The subjects of the major ceremonial cycles include origin and creation, migration, celebration of new laws, and commemoration of legendry and mythic occasions. Each serves to hold society together, create harmony, restore balance, ensure prosperity and unity, and establish right relations within the social and natural world (Allen 1996, p. 259).

The literature of a particular time period or a region can provide access to representation of nature and also help us understand the presence and absence of nature and its relationship to human beings.

2.9 Nature Conservation

The history of conservation derives from the concept of wilderness in many ways. Johnson (2007, p. 112) recounts the growth of conservation as a practice: "The turn of the twentieth century witnessed the transformations of wilderness as an idea into wilderness as practice: the creations of parks and other areas set aside from settlement". It is true that when one talks of conserving nature today, the reference is usually to the idea of conserving an ecological landscape—an area that is demarcated as a "nature zone"—called by various terms such as national park, wildlife sanctuary, nature preserve, and biosphere. The activity of nature conservation involves not only presupposes the concept of nature, but also questions the nature of wilderness, construed as exclusive of human beings. It also questions its appropriateness for indigenous communities that have lived in such areas for a long time. Conservation of resources is another significant area that has evolved in the discourse of sustainability. The limitation of certain natural resources available for the use of human beings and the rapid conversion of such resources into products for consumption has created a scenario where it is likely that many of these resources may not last for very long.

In his essay, "Science, Nature and Globalization of the environment, 1870–1990", Frank (1997) argues that the entity nature was conceptually reconstituted over the course of the last century from the conceptualisation of nature as a cornucopia of resources to a perception of nature as a universal life-sustaining "environment" or "ecosystem". Drawing on data from the various international treaties pertaining to environmental issues, he shows that there was a positive effect on the concern for the environment when the concept of nature changes from being a cornucopia of resources to the idea that it is an environment we occupy. It is possible that the idea of nature conservation became popular after this conceptual change. The impact of human beings on the environment and the other species of the planet became clearer when the results of environmental crises in the West such as the Dust Bowl Syndrome and the effects of pesticide usage received attention. The idea of sustainability is interrelated to the concepts of conservation and preservation of resources. Norton (2003) describes the perceptions of "strong and weak sustainability" which he claims are two perspectives about what is to be preserved for future generations.

He writes: "'Weak sustainability' refers to the maintenance, into the future, of a non-declining stock of aggregated capital; according to this definition, a culture is acting sustainably if each generation passes on to the next as much capital in the form of natural resources, wealth technological capabilities, labouring power, knowledge etc., as they inherited from their predecessors". Norton (2003) also suggests that in contrast, strong sustainability poses more demands than this type of requirement. It is expected under the idea of sustainability that "… each generation protect certain specified processes and features of natural systems as essential elements of their bequest to future generations" (p. 481). Norton claims that such features are referred to in terms like "health" and "integrity" that are not just ecological terms but are a part of public policy discourse. They are not only descriptive, but also evaluative (pp. 481–482). How does one measure the health of the ecosystem or its integrity? Can there be evaluative models within Indian traditions of thought that can suggest culturally relevant alternatives to some of these terms? These are some questions that will be taken up for discussion later in this book.

The belief that wilderness had to be conserved for the protection of other non-human species on the earth led to the development of Conservation Biology. Popularly called a crisis discipline or activist discipline, a major focus of this discipline is on reducing the loss of species and populations, and habitat fragmentation. Biodiversity conservation spans multiple levels of biological concepts ranging from genes to landscapes and also interactions between the different levels.

Philosophers claim that certain world views form the fundamental foundation of many ecological management practices. According to Sasidhar (2006), a transformation of these world views requires the identification of possible philosophies that can effectively reorient the current paradigms of conservative action. Soper (1995) in her book *What is Nature* also suggests that representation of nature may have political implications in the field of ecological conservation.

As an offshoot of the study of animals, ethology, or the study of animal behaviour, was a very descriptive science that slowly gained importance with the development of ecology, as it was clear that animals were deeply related to the environments they lived in. Marshall (1992) describes the split in the fundamental conceptualisations of the animal behaviour studies:

> By the First World War, ethologists had developed into two camps: the vitalists, who believed in instinctive drives, and the behaviourists, who did not go beyond describing what they saw in laboratories. But an increasing number were determined to observe animal behaviour in their natural surroundings in the wild (p. 235).

As the importance of observing animals in the wild began to be more significant than laboratory studies, a large number of popular books on animal behaviour and the interaction between animals and humans were written. Renowned among these writers are Konrad Lorenz (1903–1949, who wrote many books on animal and human behaviour) and Edward O. Wilson (who wrote *Sociobiology: The new sythesis*, in 1975) who represent some of the major streams of thought in this discipline. It is clear that without the central concepts of ecology that deal with the relations between organisms and their environments as an integrated whole in the natural world, animal behaviour studies would be incomplete (Marshall 1992).

Today, we can say that ecology has also moved away from being a discipline that has traditionally been investigating the biophysical world, towards a discipline that also concerns itself with areas of human–nature relationships. It has become imperative for these disciplines to examine the different aspects of the relationship between human beings and nature. So far, ethical frameworks have dealt with the behaviour of human beings towards other human beings or at most towards other living beings. With the development of the sciences of conservation biology and ecology, the position of the human beings and their role in the biophysical environment have necessitated a paradigmatic shift in the current ethical and value systems. Conceptualisation of nature is one of the philosophical presuppositions that can be foundational to many of the key issues ecologists and conservation biologists are facing today. Since the historical trajectory of a Western concept of nature as discussed earlier cannot be undone, many environmental philosophers are looking at other systems of thought that have escaped this transformation and trajectory While some philosophers are going back to ancient traditions of Western world in search of conceptual resources (such as the Greek concept of Gaia, or the Christian beliefs of stewardship), some others are examining conceptualisations of indigenous traditions of American Indians and Aborigines of Australia. Yet, others emphasise the study of Asian traditions of thought.

The prescriptive and moral dimensions of these traditions of thought are often supported by various metaphysical and epistemological considerations that include concepts of nature as a category or a certain idea about the ontological status of nature and its components in traditions of Indian thought—Vedic, Sāṃkhya, Sāṃkhya-Yoga. Most beliefs, conceptualisations and practices of India exist within a larger tradition of philosophy and this background cannot be ignored in any serious project in Indian thought. The final elucidation of prescriptive practices in this work attempts to address the substantive and normative practices of conservation and action.

References

Allen, P. G. (1996). The sacred hoop: A contemporary perspective. Republished in 1996. In C. Glotfelty & H. Fromm (Eds.), *Ecocritism reader: Landmarks in literary ecology*. Athens, Georgia: University of Georgia Press.

Aristotle. (Trans.) (1963). Physics book ii.1. In *The Philosophy of Aristotle, Introduction and commentary by Renford Bambrough* (A. E. Wardman & J. L. Creed, Trans.). New York: Penguin books.

Bowie, A. (2001). Friedrich Wilhelm Joseph von Schelling. In E. N. Zalta (Ed.), *The stanford encyclopedia of philosophy* (Winter ed.). Available at http://plato.stanford.edu/archives/win2010/entries/schelling/.

Brutt, E. A. (1924). *The metaphysical foundations of modern physical science*. London: Routledge and Kegan Paul ltd. (Reprint from *International library of psychology, philosophy and scientific method*, 1967).

Carone, G. R. (2001). The classical Greek tradition. In D. Jamieson (Ed.), *A companion to environmental philosophy* (pp. 67–80). Malden: Blackwell Publishing.

Collingwood, R. G. (1945). *The idea of nature*. Oxford: Clarendon Press.

Comford, F. M. (1997). *Plato's cosmology: The Timaeus of Plato*. Indianapolis: Hackett Publishing.

Frank, D. J. (1997). Science, nature and globalisation of the environment, 1870–1990. *Social Forces, 76*(2), 409–435.

Glotfelty, C. (1996). Introduction: Literary studies in the age of environmental crisis. In C. Glotfelty & H. Fromm (Eds.), *Ecocritism reader: Landmarks in literary ecology*. Athens: University of Georgia Press.

Hibben, J. G. (1910). *The philosophy of the enlightenment*. New York: Charles Scribner's Sons. http://www.questia.com/PM.qst?a=o&d=23222127. Accessed 25 May, 2004.

Howarth, W. (1996). Some principles of Ecocritism. In C. Glotfelty & H. Fromm (Eds.), *Ecocritism reader: Landmarks in literary ecology*. Athens: University of Georgia Press.

Johnson, B. (2007). Wilderness parks and their discontent. In M. Lewis (Ed.), *American wilderness: A new history*. New York: Oxford University Press.

Lafrenniere, G. F. (1985). World views and environmental ethics. Special Issue: Roots of ecological thought. *Environmental Review*, *9*(4), (Winter).

Leiss, W. (1994). *The domination of nature*. Montreal: McGill–Queen's University Press. http://www.questia.com/PM.qst?a=o&d=80437339. (Stable URL).

Love, G. (1996). Revaluing nature: toward an ecological criticism. Republished in 1996. In C. Glotfelty & H. Fromm (Eds.), *Ecocritism reader: Landmarks in literary ecology*. Athens: University of Georgia Press.
Lovejoy, A. O. (1927). 'Nature' as aesthetic norm. *Modern Language Notes, 42*(7), 445–451.
Marshall, P. (1992). *Nature's web: Rethinking our place on Earth* (1st ed.). London: Simon and Schuster Ltd.
McClure, M. T. (1933). The Greek conception of nature. In *Proceedings and Addresses of the American Philosophical Association* (Vol. 7, p. 120).
Merchant, C. (2004). *Reinventing Eden: The fate of nature in western culture*. NewYork: Routledge.
Naddaf, G. (2006). *The Greek concept of nature*. New York: State University of New York Press.
Norton, B. G. (2003). *Searching for sustainability: Interdisciplinary essays in the philosophy of conservation biology*. Cambridge: Cambridge University Press.
Sasidhar, K. (2006). Ecology and vedic vision. In D. P. Chattopadhayaya (Series Ed.) & R. Balasubramanian (Vol Ed.), *The enworlded subjectivity: Its three worlds and beyond, philosophy and culture in Indian Civilization* (Vol. 11., pp. 697–724). New Delhi: PHISPC, Centre for Studies in Civilizations.
Soper, K. (1995). *What is nature?: Culture, politics and the non-human*. Oxford: Blackwell Publishers Ltd.
Tellenbach, H., & Kimura, B. (1989). The Japanese concept of nature. In J. B. Callicott & R. T Ames (Eds.), Nature in Asian traditions of thought. Albany: State University of New York Press.
Vogel, S. (2005). Nature as origin and difference: On environmental philosophy and continental thought. In M. E. Zimmerman, J. B. Callicott, K. J. Warren, I. J. Klaver, & J. Clark (Eds.), *Environmental philosophy: From animal rights to radical ecology*, 1993 (4th ed., pp. 296–309). New Jersey: Pearson Prentice Hall.
Wood, D. (2005). What is Ecophenomenology? In M. E. Zimmerman, J. B. Callicott, K. J. Warren, I. J. Klaver, J. Clark (Eds.), *Environmental philosophy: from animal rights to radical ecology*, 1993 (4th ed., pp. 311–325). New Jersey: Pearson Prentice Hall.

Chapter 3
Conceptualisations of Nature in Indian Traditions of Thought

Abstract This chapter analyses the equivalents of the term "nature" in Indian philosophy. After an initial survey of various concepts related to nature, I discuss in some detail a few terms like *prakṛti*, *nisarga*, *padārtha,* and *dravya* from different schools of Indian philosophy. The process of evolution and creation of the material world and the way things are related to each other is described. As some of these terms persist in the contemporary times as equivalents of nature, I dwell at length on some pre-modern interpretations of these concepts and also briefly suggest the ecological implications of such understandings.

Keywords Translation · *Loka* · *Prakṛti* · *Nisarga* · Nature in Indian philosophies · Categories · *Padārtha* · *Dravya*

3.1 Conceptualisations of Nature in India: Available Sources

Religious and cultural world views have often shaped our relationship and our conduct towards our environment. Gottlieb (2004, p. 9) writes of the influence of religious traditions on the human attitude to nature, "In short, religions have neither been simple agents of environmental domination nor unmixed repositories of ecological wisdom. In complex and variable ways, they have been both".

When we see the presence of sacred groves or trees in rural India, we can infer that cosmologies continue to influence the present-day narratives of social and environmental concerns. Particularly in the Indian context, the influence of traditions and textual narratives associated with religious practices and rituals in our attitude to nature cannot be ignored. Singh (2005, p. 106) writing about the Vedas in particular claims that "Ancient Indian texts are a rich source of information and insight on the historical roots of Indian environmentalism". Others like Narayanan (2001, p. 182) have also emphasised the importance of textual and other resources

M. Baindur, *Nature in Indian Philosophy and Cultural Traditions*,
Sophia Studies in Cross-cultural Philosophy of Traditions and Cultures 12,
DOI 10.1007/978-81-322-2358-0_3

for environmental ethics from Hindu thought: "The resources from which the Hindu traditions can draw in approaching environmental problems are several and diverse: there are texts, of course, but also temples and teachers".

The categories and terms used to describe the universe and its components and experience of the world in Indian thought are distinct from Western philosophical conceptualisations of nature as "non-human". The dichotomy of human beings versus the rest of nature is not a rigid value designator either. Rather, one could say that the conceptualisation is based on a certain world view of "beings and the metaphysical worlds" they live in. I refer to this as a *loka*-centric view of the universe. According to Indian cosmology, the cosmos consists of many worlds called *loka*s (which are sometimes mythical and not *on* the earth) and also beings (*bhūta*s) that inhabit such worlds.[1] Kinsley (1995, p. 55) elaborates on this perception, when he explains "In many Hindu scriptures, it is clear that the world is perceived as being alive with forces, powers, spirits, and deities that express themselves through what we call natural phenomena". All elements of a cosmos find a place in a complex hierarchical humanised cosmic order in Indian thought.

In Indian thought, though there were many pre-modern schools that had similar metaphysical and ethical positions, they do not have as much prominence as the dominant forms of philosophical systems—the Vedantic schools of thought, such as non-dualism, dualism, and special dualism. Some of these concepts continued to flourish and were embedded in other traditions of knowledge such as Āyurveda or Tantra. Therefore, the resources for environmental ethics in Indian philosophy will have to be drawn from a set of pre-modern traditions that are to be found in a context that is historically earlier to the awareness of an environmental crisis.

As we know that rituals and traditions influenced by such conceptions continue to evolve in practice along with discourses of nature that are perpetrated by mythology as well as the narratives of the divine, it is possible to at least undertake a study of nature in Indian philosophy. Not all writers are convinced that there are equivalent terms for nature in an Indian context. Summarising the idea of nature in some streams of Asian thought, Barnhart (1997) suggests that it is appropriate to look for a family resemblance in the varied notions of nature. He suggests that these streams of thought are clustered around an idea of reality that is different from Western thought: "I would suggest that the dominant distinction in these Asian traditions is not body/mind but relative versus of absolute beings or Being" (p. 424). His key point is that the main interpretation that differentiates the Western concept of nature from Indian conceptualisations is that the idea of nature is defined by its exclusion of "non-human or non-artificial". If the concept of Western nature were to lose this sense, then conceptualisation of the Eastern and Western notions of nature would be more of a "family resemblance".

[1]For details on some of these sacred beliefs, see Chap. 7 of this book on the idea of nature landscapes as sacred.

3.2 Problems of Translation

What are some of the philosophical terms related to the concept of nature in Indian thought? The concept of nature in Western thought has evolved within particular historical and cultural contexts. Therefore, it is unlikely that we would find a completely equivalent term for the exact word "nature" in Indian traditions of thought.

Mohanty (1999, p. 220) suggests that four trends somewhat similar to Greek thought evolved out of a search for the ultimate element in Indian thought. His arguments are summarised below.

1. The "conception of natural law", which took two forms the imposed law "came to be the law of the unseen (*adṛṣṭa*)" and immanent, "as the real source of unity underlying the plurality of appearances, led to the idea of *brahman*, the *ekam sat*, the one existent (another transformation of the Vedic *ṛta*)..." (p. 220)
2. The atomic theory of the Vaiśeṣikas, which is different from Greek atomism. He writes "The Indian atomists, however, stopped with qualitative atoms (the earth-atom, the air-atom, the fire-atom, and the water-atom), each with its distinctive simple quality" (p. 220).
3. The essentialism found in some form in the Nyāya-Vaiṣeśika school.
4. A teleological interpretation that forms a part of the Saṃkhya philosophy.

Before I begin to describe and discuss some of these conceptualisations, I must make some preliminary observations about the problems of translation.[2]

As seen in Chap. 1, we have noted that the word "nature" within the English language has carried many meanings across time and cultures. We must agree with Bruun and Kalland (1995, p. 9) who discuss this complexity: "In discussing nature we are dealing with maybe the most complex concept in western languages that in common usage has a complicated repertoire of meanings". They also add that the concept of nature is in a state of transition even in the current times.

One can say that while the nominal component of "nature" has remained the same, the conceptual component has varied again and again. Firstly, it is clear that in most translations, terms that refer to concepts are treated nominally, like the "name" of some abstract idea which is captured as accurately as possible. It is also believed that using a term in a different language "to name" a particular concept would serve as a translation (Sarukkai 2002). For instance, the word "*prakṛti*" is used to name the concept "nature" as it is used in the context of Western thought. In places where one would use the "concept nature", such as in the phrase "nature conservation", the term "nature" in the Indian languages (like Hindi and its derivatives) is replaced with *prakṛti*. So can we just say that *prakṛti* is nature?

[2]I am aware here that the idea is more about translation of "concepts" which can be regarded under the same theories and philosophy of translation. Many of these ideas presented here are a result of my discussion with Dr. Sundar Sarukkai on his paper "Translating Concepts: some issues in the methodology of history of non-Western science" (Sarukkai 2010).

This would be using a historically pre-modern term to refer to a modern concept or idea. This, however, is not the only the issue around equivalence.

Bruun and Kalland (1995) aptly summarise the complicated project of investigating nature as a concept both in Eastern and Western thoughts. They point out that even within a single tradition of thought in the Western tradition nature has references to the physical world, and at the same time, it refers to an abstract space containing many principles and ideas. It also comes attached to a number of other concepts such as "environment" or "landscape" or even the concept of qualities or "properties". The conceptualisation of nature in Indian thought is complex. To examine the idea of "nature" within Indian traditions of thought involves a difficult process of translation, even if the main purpose is to merely describe it. As Sarukkai (2002) notes, within a naïve view of translation, it is believed that somehow in the process of translation, "the essence" remains intact. He points out that any kind of translation—intralingual, interlingual, or intersemiotic—raises the main problem of translation—that of incommensurability. After a discussion on these types of translation, he summarises that complete equivalence is almost impossible to achieve in any translation. And he notes "The idea of complete equivalence, as also the notion of faithful translation, has been the bane but also the stimulus towards generating more complex theories of translation". He adds that any primary impulse that inspires an act of translation responds to "an original" that is an "already given".

Extending this to conceptual translation, I consider the idea of "nature" from Western thought as the original concept for my work.[3] Though not engaging in a complete comparative study of the two traditions of thought, I engage with Indian conceptual thought with a background reference to the Western concept of nature and the possibility of an ecological ethics. I insist that the very problem of translation becomes an advantage in this project. The fact that the concept of "*prakṛti*" is not equal in all respects to the concept of "nature" allows us to develop an alternative representation of nature itself. Instead of similarities in the conceptualisations, examining the diversity here will allow for emergent derivations of environmental ethics. Take for example the word *nisarga* that is sometimes used to represent the concept of nature. While nature derives from the idea of "the way things are", or from the idea of "an entity opposed to human beings" originally, in Sanskrit, *nisarga* derives its meaning from the idea of creation.

Barnhart (1997) also raises an interesting point about the differences in meaning of "nature" and the "equivalents of nature" in Asian context. He claims that Western philosophy has been engaged in the task of reconciling the two different meanings of nature: the first which refers to "not artificial, free of human contrivances" and the second meaning which carries a sense of "the way something is, its *physis*". He argues rightly that the equivalents of nature in Asian thought are closer to the second meaning and therefore could lead to different paradigms of ethics in environmental philosophy.

[3]From a discussion with Dr. Sundar Sarukkai in February 2010.

Despite the differences in interpretations, we are sure that the terms related to nature in the Indian traditions have certainly existed and evolved within different traditions of thought, thereby providing us with a rich source of conceptual categories to understand the world around us. Among the terms equivalent to nature in Indian thought, there are two types. One set of such terms carries the original metaphysically relevant meaning of nature, similar to that of "*physis*" in Greek or "*natura*" in Latin. The second set of terms are related to a broader sense of nature, conceptually different but referring to the sense of the natural world or parts of the environment. Indian philosophical thought has a wide spectrum of theories about the world ranging from realist to the extremely idealist schools. In simpler terms, some philosophies like Nyāya insist on the metaphysical reality of the world and its various components. On the other hand, some like the Advaita school believe that the world is only a false projection of some ultimate reality. However, in their own spheres, each of these philosophies has something to contribute to the idea of nature.

From an initial survey of the literature related to nature, I found the terms *prakṛti* and *nisarga* in Sanskrit are related to both interpretations of the term nature. These terms show many instances of synchronic usage to the word "nature" in textual traditions.[4] It is also important to note that there is no strict one-to-one correspondence of any of these terms with the term nature nor are the textual instantiations of these terms strictly equivalent to the current semiotic usages of these words in common Indian languages. It would be, however, prudent to say that these terms are deeply related to the conceptualisation of nature in Indian thought. The two terms—*prakṛti* and *nisarga*—used for nature in many Indian languages hold symbolic and semiotic meanings derived from their philosophical usages and therefore lend themselves to somewhat richer conceptualisations of nature for purposes of understanding the relationship between human beings and nature.

One significant point to keep in mind here is that these types of conceptualisations are linked to a system of metaphysics within a larger school of Indian philosophical thought and they have to be understood within that context. On the other hand, the problems with such conceptualisations are the poor applicability of such philosophical ponderings on metaphysics to the real-world issues of nature. Rolston (1987) stresses that the idea of nature has to be such that there is a "translating scheme" of sorts between the categories of a vision of nature that informs conceptually and the practical on the ground decisions related to conservation:

> A metaphysics can hardly be expected to provide a blueprint for action, but if a metaphysics cannot orient action in some meaningful way, then it is of no help where the West needs help—valuing the environment that humans inhabit. Such a theory cannot be put into practice environmentally, though perhaps it can be put into practice in other ways—existentially or soteriologically (p. 186).

---

[4]Detailed examples of such usages will be provided in the following sections of this chapter.

He insists that alternative metaphysics may help to counter the metaphysics that gave rise to modern science, creating a "metascience" understanding of the world that may not necessarily contradict science, but instead provide us with an alternate logical system to understand nature (pp. 182–184). There will be problems in trying to adapt Eastern metaphysics to Eurocentric sciences. But conceptually analysing nature will need us to explore all possible streams of thought to lay a foundation perhaps for future work.

One could categorise the conceptualisations of nature in Indian philosophy into three types. The first type includes those terms and equivalents of nature that are concerned with compositionality of the universe or cosmos at its most fundamental unit of constitution. Nature in this kind of conceptualisation is like the essence or basic building block of a cosmos or creation.

The second type of conceptualisations is based on the idea of nature as the phenomenal whole, the experience of creation itself. The idea of nature as the cosmos and the universe is common enough in many pre-modern traditions. In Indian, thought creation is referred to as *sarga*. The word "*nisarga*" etymologically means "downward" evolution or creation.[5] As a prefix, "*ni*" usually refers to a negation of the word. The word *sarga* clearly refers to creation. So then, *nisarga* would mean the same as non-created or self-created. This, however, is in opposition to many passages in the Vedas and other texts that clearly indicate the world is "created" and has evolved downwards from a supreme being. The prefix "*ni*" could also mean downwards.[6] This term can be interpreted from almost all cosmologies of creation described in Indian traditions that see evolution from a primordial being or cause.

Yet a third type of conceptualisation can be said to be more cultural than philosophical. Nature is understood through a tradition of narratives and practices as divine, or feminine, as non-human or as sacred landscape. In other forms of interpretations, nature also represents place and geography of the terrestrial. Nature as described by Literature and other cultural traditions also falls under this set of ideas.

The evolution of this universe finds its mention in Vedas, and most classical philosophies of Indian thought insist that what is found in the Veda is valid knowledge and must be accepted as authority. However, the different schools of thought and philosophers interpret the statements of the Vedas in different ways and are also free to use their own perspectives to understand the conceptual insights from these texts.

The question in the study of a pre-modern philosophy of nature can also be metaphysically interpreted. We could for instance ask of the Indian philosophical system, what kinds of things are there in this creation and how are they related to each other. Mohanty (1999, p. 205) lists the two main ways of theorising change and permanence in Indian thought:

---

[5]This meaning was provided by Sri Siddharth Arya, one of the few traditional *Sāṁkhya* scholars of the *gurukul* system in India.

[6]Like in verbs such as *nipatati*—falls down, and *niṣati*—sits down.

> The problems of continuity versus discontinuity and change versus permanence concerned the Indian thinkers as much as they did the European thinkers. Hindu thinking took two forms: the atomism of the *Vaiśeṣikas* and the theory of original, *Ur*-Nature (*Prakṛti*) of *Sāṁkhya*. The former found discontinuity at the root of things, the latter continuity. The former saw in the last analysis unchanging atoms, the latter understood the ***Ur***-Nature as an incessant becoming—homogeneous at first, of like into like (*sadṛśapariṇāma*), and only subsequently heterogeneous, i.e. of like into unlike (*visadṛśapariṇāma*).[7]

Given the premise that the category of human–non-human is not very distinctive, one could analyse nature in Indian philosophy through a comparison of the existents in different schools of thought. In the following sections of this chapter, some key thematic positions of the metaphysics of nature are examined. The first is the conceptualisation of nature as constituted by some fundamental things. Similar to some of the Greek philosophers, the Nyāya-Vaiṣeśika school answers the question "what is nature?" by describing what kinds of things are there in nature. They also posit constituents of reality at the most fundamental and irreducible levels in terms of a theory of atomism and categories. Then, there is the discussion about the original nature, *prakṛti* from the school of Sāṁkhya, which is sometimes called a teleological philosophy. This is similar to conceptualisations of nature as origin. Finally, a third alternative would examine nature as a god-created reality with a variation from the usual theological arguments of the West. This panentheistic tradition of *viśiṣṭādvaita philosophy* is one of the significant Vedanta schools that also underlies the practices of a religious sect in the south of India.

3.3 Nature as Categories: Nyāya-Vaiśeṣika Classifications

One could refer to the Nyāya-Vaiśeṣika school as two independent schools of philosophical thought, Nyāya and Vaiśeṣika, but it has been agreed by scholars that both schools have been syncretic and supportive of each other's claim (Potter and Bhattacharya 1993, p. 12). In this section, I draw upon texts of both traditions where some of the fundamental principles are agreed upon by both schools.

According to Potter and Bhattacharya (1993, p. 48), the reality described by Nyāya-Vaiśeṣika consists of substrata (*dharmin*), their properties (*dharma*), and the relations (*saṃbandha*) between them. Nyāya-Vaiśeṣika lists fundamental entities of the reality called *padārtha* which can be loosely translated as category. For Nyāya-Vaiśeṣika, the substrata and their properties are different entities listed in separate categories.

Accordingly, the world is made up fundamental categories or labelled constituents called *padārtha*. The word *padārtha* can be translated as the "meaning of a word", while another meaning could be purpose or resource. Mohanty (2000) translates *padārtha* within the context of the Indian theory of meaning as "what is referred to by words". The Nyāya-Vaiśeṣika school offers one of the most

---

[7]Emphasis in the original quotation.

descriptive accounts of a realist world categorised into knowable and nameable existents within the context of a systematic knowledge. The categories include ontological lists of kinds and concepts. These lists seem to be descriptive, but it is not clear as to the exact methods that philosophers used to select them (p. 57). But it is clear that for Kaṇāda, the Vaiśeṣika philosopher, these are objects of valid knowledge (Sinha 2006, pp. 339–355). Nyāya extends the number of categories to one more than the original six, the seventh being *absence*.

One could posit that these categories are the basic fundamental units of the created world, or *nisarga*. So nature for Nyāya-Vaiśeṣika (NV), inclusive of the human being, is constituted by *padārtha*. The following is a list of *padārthas* from Vaiśeṣika texts as translated in Potter and Bhattacharyya (1993, p. 49):

1. Substance
2. Quality
3. Motion
4. Universal
5. Individuator
6. Inherence

Kanāda, the main philosopher of the Vaiśeṣika school, puts forward the theory of the physical world indicating that all beings (*bhūta*) and all things derived from what "has come to be" (*bhautika*) can be listed under three dimensions of reality. These are matter in the form of eternal elements (*nityadravyāṇi*), organic bodies (*śarīra*), and the internal organs (*indriya, manas,* and the self).

*Dravya*s are one of the nine irreducible *padārthas* (categories) listed above. Dasgupta (1922, pp. 3, 4) translates this term as "thing". Sometimes, this is also translated as "substance". Mohanty (2000, p. 44) also translates this as "thing" and defines the concept as

> (1) The locus of qualities (*guṇa*s) and actions (*karmas*) (2) as the material cause (or inherence cause) of the effects (i.e., that in which effect inheres).

The three—*dravya*, *guṇa*, as property, and *karma*—together form a central theme of the NV universe. Substance, or *dravya*, is derived from the basic relationship of locus and what is located there (Potter 1977, p. 69). This is linked to the other two categories, quality (*guṇa*) and action (*karma*), through the relationship of inherence (*samavāya*). Kaṇāda describes *dravya lakśaṇam*—the distinctive features of *dravya*—as a locus for actions and qualities (*kriyāguṇavat*). It also acts as a basis for its own causality (*samayāyi karanā*). As we can see from its definition, any *dravya* (thing) is potentially the locus of any *guṇa* or *karma*, but the non-*dravya* cannot be the locus of anything. Mohanty (2000, p. 44) states that "Qualities and actions and universals inhere in the *dravya*. There are *dravyas* that inhere in other *dravyas* that have parts". *Dravya*, thus, is the only category that can be the locus of anything, including other *dravya*s that may be its parts. There are fundamental *dravya*, those that are partless and composite *dravya*s that have parts.

There are nine kinds of *dravya*s—the five elements, *dik* (directional space), *kāla* (time) *ātman* (soul), and *manas* (mind). The NV philosophers claim that all things are reducible to these fundamental types. The five material *dravyas* are the four elements and the internal organs. The immaterial *dravyas* are four—time, directional space, space, and the self.

The *dravyas* are also divided sizewise into three categories—atomic, middle sized, and those that are ubiquitous (omnipresent) (Potter 1977, p. 74). The atom (*aṇu*) is minimum sized, partless, and cannot be apprehended by the senses (*Nyāya Sūtra* 4.2.16). The middle-sized ones are different composites that are perceptible, and space is a kind of the third category. The first four of the elements are earth, water, air, and fire and are permanent in their non-composite forms as atoms, but in their composite form, they are impermanent and can be destroyed. The others including the space are eternal.[8] The Vaiśeṣika philosophers posit the formation of the *dravyas* through their theory of atomism:

> Two atoms combine, on the *Vaiśeṣika* theory, to form a binary atom (*dyaṇuka*), and three of the latter sort combine to form a tertiary atom (really consisting of six atoms) or *tryaṇuka*, the smallest perceptible entity (identified with the dust particles one sees streaming along the ray of light through an opening). Out of different combinations of these arise the sundry empirical objects such as sticks and stones (Mohanty 1999, p. 212).

Potter (1977, p. 52) suggests that *dravya* can be understood through the relationship of contact called *saṃyoga* "… (Contact) is capable of relating two substances at least one of which is material (*mūrta*). Contact inheres in the pair of substances". The element space in the NV philosophy does not inhere in other *dravyas*, nor do other *dravyas* inhere in it. This is because space is partless (which is a condition for it to be eternal).

The condition that all things are in contact with *ākaśa* leads us to an interesting hypothesis. If contact has to inhere in a pair of substances, at least one of them has to be material (*mūrta*). *Ākaśa* itself is non-material. So it follows all things are material in any pair of entities in contact. This seems to be another way of stating that things occupy space, but for NV it is important to understand that space is also a thing, a *dravya,* and not an accommodating void. The whole–part relationship in NV philosophy is discussed as the *avayava–avayavin* relationship. The parts of a whole "inhere" in the whole, by this relationship. For NV, the whole is greater than the sum of its parts. Potter (1977) clarifies the position that, "In Nyāya-Vaiśeṣika a whole is produced from its parts but is not constituted by them" (p. 74). The NV philosophers stand in opposition to the doctrine of *satkāryavāda* of the Sāṃkhya philosophers. The whole is a completely new thing produced from parts. This is called *ārambhavāda*:

> According to the Vaiśeṣika and Nyāya, this same principle applies not only to the field of language, but to the world in general: What is truly eternal, such as the ultimate substances and the universals, does not change at all; and where there is change, there must be production, destruction, and replacement (Halbfass 1992, p. 188).

[8]These elements will be discussed in detail in Chap. 4.

What one finds interesting in the explanation of these categories are the relations between them. In a world struggling to define the so-called web of relations between natural elements and human beings in the current times, one could refer back to the sophisticated technical terminology that the Nyāya philosophers use to describe such relations. Potter and Bhattacharyya (1993, pp. 49–50) point to two of these important relationships. One relationship is called *āśrayā-śritasambandha*, the support-dependent relationship.[9] This implies that two of these fundamentals are related in such a way that the resider resides in, or is dependent upon the residence. There is another type of relationship that the Nyāya philosopher posits, such as quality of colour which resides or is sheltered in/by substance. In such a case, the relationship indicates an unequal importance to the support, without which it seems the dependent has no existence. The second kind of relationship is where there are qualifier–qualificand relationships called *viśeṣaṇa-viśeṣyasambandha*. In the second set of relations, directionality is implied and also it seems to be more of an epistemological category (Potter and Bhattacharyya 1993, p. 50). The Nyāya philosophers posit a special relationship to connect substance qualities and different pairs of fundamental substances called inherence—*samavāya*. Inherence technically can be defined as a relationship between two inseparable things as located to the locus. One would say inherence holds the position of glue or a keeping-together force. Accordingly, (from Potter and Bhattacharyya 1993),

> Inherence relates qualities, motions, universals, and individuators to substances. It also relates universals to qualities and universals to motions. Finally it relates composite individuals to the "parts" which are the cause of composite individual (p. 51).

The importance of inherence as relationship that is "marked through knowledge" (p. 51) indicates that inherence is dependent on a subject attending to the relationship between, say, a colour and the substance. Further, inherence is related to its relata through a relation of identity, called *tādātmya* or sameness. That means that inherence is independent of its relata, in some way called into being from potency when two individuals are related.

The idea of inherence is very interesting to postulate as a way of "seeing relationships" when we see two related objects. Take for instance the concept of nature as constituted by many objects that are said to be its parts.[10] Here, I am only trying to interpret the idea of nature that can be construed as the set of all objects, processes, and relationships that constitute our environment. Drawing upon earlier described conceptualisation of nature and from a realist position, one could argue that these relations exist independent of the human subject's cognitive attention. If we for the time being jeopardise the realist position accorded to Nyāya and admit

[9]The term used by Potter and Bhattacharyya (1993) is resider–residence, but since the primitive locus of the Nyāya is a loci relationship that can be defined by "in" or "at" or "on", the translation "residence" seems to be biased to view of the loci as a container that I want to avoid. The possibilities of shelter are not spatiotemporal or related to materiality in this case.

[10]In philosophical context, the relation between whole and parts is a much debated subject and to refer to nature as whole constituted by parts is also problematic.

as per the texts the Nyāya claim that inherence is subject to knowability and namebility, then we can hypothesise that the inherence relations in nature require attention of the human being. So it is "perception" of relationships between nature and its constituents that makes us "know" that something is a part of nature.[11] Even if the absolute metaphysical reality does not validate the idea that nature is whole, the fact that we are able to see—let us say a category we call forest—as constituting a part of a larger category of nature, one could say inherence relates the two. The category forest is sheltered by the nature that is the shelterer. Since the relation of inherence does not require that both relata are tangible and substances, it is also possible to relate qualities and create ecological relations through invoking Nyāya categories.

The above discussion shows interesting possibilities of invoking alternate metaphysics in the service of contemporary ecological concepts. The category of sheltered and the shelterer is also present in a different way in Sāṃkhya philosophy that will be taken up for discussion in a later chapter in detail.

3.4 *Prakṛti*: A Survey of Meanings

The sanskritised word for nature in many Indian languages is *prakṛti*, derived from Sanskrit. "Nature" is translated into *Prakṛti* in many Indian languages. The reverse translation of *Prakṛti* into English word nature is problematic. *Prakṛti* is nature—in the sense that it is the source and power from which things are produced—not nature in the sense of an extra-human world that the word nature signifies. *Prakṛti* in Sanskrit particularly is also used as a technical term in the Vedas. It also refers to the primary sounds, stems of words, and primary sacrifices from which multiple modifications (*vikṛti*) are derived. In rituals, *prakṛti* means model rituals or archetypal sacrifices from which variations can be derived.

Jacobsen (2002) traces three clusters of meanings for the word *prakṛti*. Besides its technical reference within *Sāṃkhya* philosophy,[12] the word *prakṛti* is often used in many texts such as those of drama, mythology, phonetics, or grammar. Jacobsen (2002) describes these three clusters of meanings for this term:

1. *Prakṛti* is "that *which precedes*", "*first*", "*that which is in its own form*"; therefore, it is used in contexts like natural, archetype, one's essential character, and normal.
2. *Prakṛti* in plural refers to "*components, constituents, and the parts*" of a whole such as a human being or the political state.

---

[11]Valid knowledge called *prama* can be obtained by direct perception according to almost all Indian philosophies.

[12]*Prakṛti* forms a part of the dyad the material principle that coexists with the Consciousness principle (*Puruṣa*). See the following Sect. 3.5 on *Sāṃkhya* philosophy for more details.

3. *Prakṛti* also refers to "*material cause*", "*producer of effects*", and "*innate power of transformation and manifestation*"; here, *prakṛti* gets associated with the field of production and in later periods is associated with women and goddesses.

In *Sāṃkhya*, there are two ontological categories, the sentient *puruṣa* and the insentient "matter" principle called *prakṛti*. These concepts were engendered, and *prakṛti* came to represent the female principle. For instance, Monier-Williams (1990) describe *Prakṛti* as a goddess who is the material cause of the world and all created things. She is the prototype of the female sex. This idea is also directly linked to a perception of the earth as a goddess. Bhattacharya (1982) suggests "In the present form of *Sāṃkhya*, as well as the *Tantras*, the term *Prakṛti* has acquired a purely metaphysical connotation, but basically it stood for the mother Earth, the fruit-bearing soil".

In modern Hindi, the word *prakṛti* is used to mean nature and the term *paryāvaran* to refer to the environment.[13] Another word that is often used in the *Vedas* to mean "surrounding" is *parisara*. It gets its meaning from the prefix *pari*, "to surround".[14]

The opposite term of *prakṛti* as referring to the primary is the term *vikṛti* which means secondary or derivatives (Jacobsen 2002). Sometimes, *prakṛti* is considered to be the primordial unmanifest nature and manifest nature is also referred to as *vikṛti* (Rao 1963). The term "*vikṛti*" in the *Sāṃkhya Kārikā* is used in the meaning of an "evolute": "Primordial nature (*Mūla-prakṛti*) is uncreated. The seven (derivatives)—*mahat*, etc.—are both evolvent (*prakṛti*) and evolutes (*vikṛti*)…" (*Sāṃkhya Kārika* 3, henceforth to be referred to as *SK*) [15] Evolutes are considered to be effects of a primary cause, sharing the materiality but different in terms of being modified.

An important aspect of the concept *Prakṛti* is its oneness as the source of the evolution of its variants. Jacobsen (2002) writes "… the oneness of matter is a presupposition for the empirical world to be a common field of experience" (p. 231). *Prakṛti* evolves into an infinite variety of related evolutes that makes up this phenomenal world, yet maintaining the oneness. The Sāṃkhya school of thought (as well as its subsequent derivative, Sāṃkhya-Yoga philosophy) perhaps offers one of the most comprehensive and clear understandings of *prakṛti* as a philosophical concept.

---

[13]For example, *prakritic soundarya* (in Hindi)—natural beauty, etc. The word *paryāvaran* (in Hindi) seems to indicate "environment" word in recent times. It means covered around and is very similar to the etymology of "environs"—"around," from *en-* "in" + *viron* "circle, circuit," from (French) *virer* "to turn".

[14]The prefix "*pari*" refers to complete, entire. From S. v "Pari" in Monier-Williams, Leumann, and Cappeller, *A Sanskrit English Dictionary, Parisara* is used to refer to nature in the Kannada language.

[15]For the original verse, see "Sāṁkhya Kārika, Verse 3", trans. Sinha (1979, II ed. Reprint, p. 4).

3.5 *Prakṛti* in *Sāṃkhya* Philosophy

Prakṛti is an equivalent for nature associated with Sāṃkhya and Sāṃkhya-Yoga schools of thought. Both are dualistic systems of thought. According to Sāṃkhya philosophy, *prakṛti* is a part of a dyad, a creation component that functions both for the enjoyment and salvation of *puruṣa*, the conscious principle. The term *prakṛti* is broader than the term nature. The entire materiality of the cosmos is said to have evolved from a primordial cause often referred to as *mūla-prakṛti*.[16] The idea that nature as *prakṛti* is the primordial origin brings into focus relatedness between all created entities based on common origin. As mentioned earlier, it is possible to imagine how *prakṛti* comes to be represented as the mother goddess in later traditions. The goddess Durga is often called *jaganmāta*, the mother of the worlds or *ādi śakti*, the primordial energy.

As origin, nature in Sāṃkhya precedes the components of creation, such as the five elements. In Sāṃkhya, the manifest world is a series of material effects from a primordial material cause. In other words, *prakṛti* is different from the five elements in the natural world that are referred to *pañcha mahābhūtā*.[17] These gross materials are evolutes of *prakṛti* and form the gross stuff—*pañchabhūtās*—called matter principles, by interactions with one another.

Also central to Sāṃkhya philosophy is the theory of causation called *sat-kārya vāda*. This is summarised by Larson and Bhattacharya (1987) as "a tripartite process of emergence that is both logical and natural". The *SK* states "The effect (*kārya*) resides (*satkārya*) in the cause (*hetu*) in a potential state prior to the operation of the cause" (*SK* quoted in Larson and Bhattacharya 1987, p. 153). There are three realms of creation in nature—*Tattva, Bhāva,* and *Bhūta*. The *Tattva* is the ontological realm, *Bhāva* is the epistemological realm, and *Bhūta* sarga or beings realm of creation corresponds to the phenomenal world of perception (*SK* quoted in Larson and Bhattacharya 1987).

In this philosophy, *prakṛti* is also used in the sense of being the ultimate first principle that all psycho-physical creation evolves from. The entire manifest world is pervaded by the "first" primordial nature (*mūla-prakṛti*), which has three coconstituting principles called *guṇas*. The *guṇas* and the *pañcha mahābhūtā* are invoked to explain the diversity of beings in the world. The world in all its varieties in the three realms is accepted as real by the *Sāṃkhya* philosophers. The oneness of *prakṛti* as primordial material is unlike the substantive oneness of *Brahman* in Vedanta philosophy. Ghosh (1977) speaks of this distinction and describes the need for Sāṃkhya philosophers to explain the diversity: "Sankya (*Sāṃkhya*) parts company, therefore with monism and undertakes a critical examination of its detail in order to discover what else is needed for their explanation".

---

[16]Material does not refer to the gross matter alone but encompasses all created "stuff" of the universe.

[17]These elements are earth, water, fire, air, and ethereal space. See Chap. 4 for a discussion on the five elements.

While it is easy to explain the insentient components of the world as different permutations and combinations of the five great elements, explaining the varieties of life forms needs a different hypothesis. The Sāṃkhya philosophers make an argument for the diversity and multiplicity of the sentient universe by invoking the idea of *guṇa*s that coconstitute *prakṛti*. Dasgupta (1922) translates *guṇa*s as "feeling substances". The cosmos which arises from *prakṛti* consists of both affective and material stuff. "The characterisation of these guṇas is in terms of psychic states or affective conditions but these are much more than that" (Larson 1979, p. 162). *Guṇa* is often translated as "quality", but the Sāṃkhya meaning is very clear in its formulation (Rao 1963): "*Sattva*, etc. are substances, and not *guṇas* in the *Vaiśeṣika* sense of the word".[18]

The commentator Vijnāna Bhikṣu describes the meanings of the word "*guṇa*" thus as follows:

In this, Sāṃkhya, Śastra, and in the Veda, etc., the word *guṇa* is employed to denote them (sattva, etc.) because they exist only to serve the ends of the *puruṣa* (and are, therefore, of secondary importance), and also because they form the cord (as it were), namely *mahat*, which essentially consist of the three *guṇas*, and which bind the brute beast (so as to speak), *puruṣa*.[19]

By this definition, *guṇa* is the "rope" that binds the *puruṣa*. The meaning of *guṇa* according to the *SK* is described by Larson (1979) thus as follows:

> … According to the *Kārika*, the *guṇas* include two levels of meaning: [1] as psychic or moral conditions—that is, *sattva* as pleasure, goodness etc.; Rajas as pain, passion, etc., And *Tamas* as indifference, dullness; and [2] as factors involved in the unmanifest and the manifest world—that is, *Sattva* as a illumination, thought, etc.; *Rajas* as activation, energy etc.; And *Tamas* as heaviness etc.

Ghosh (1977) explains that that the coherence of the three *guṇa*s in different proportions results in the entire world of creation in Sāṃkhya. And he writes "perceptibility (*sattva*), mutability (*rajas*) and inertness (*tamas*) explain every form of knowable existence or appearance". The *SK* also mentions the interactions of the *guṇa*s, "the three *guṇa*s mutually dominate, support, activate and interact with one another to produce the world" (*SK* 12). The implication of *guṇas* being "feeling substances" is that they induce values in objects that may constitute in the world. All of the world is therefore subjected to predominance of any one of the *guṇas* and is judged by the occurrence of these *guṇa*s.

According to *SK*,[20] the *guṇa*s are experienced as *priti* (agreeable), *apriti* (disagreeable), and *viśāda* (oppressive). The commentary for verse 12 of the *Sāṃkhya Kārika* suggests that each of the *guṇas* can be the cause for each other. The commentary illustrates it with an example of how a king acting as a protector

[18]The Vaiśeṣika reference is in sense or quality, or property of something. See Narain (1961), for an interesting discussion on *guṇa*.

[19]See Vijnāna Bhikṣu's commentary of "*Sāṁkhya Pravachana-Sūtram*, Book I, Sūtra 61" in Sinha (1979, p. 95).

[20]Summary and translation by Potter and Larson, in Larson and Bhattacharya (1987, p. 154).

(like *rajas*–action–cause) who brings pleasure to the good people (*sattva* results) yet can still cause mortification and violence to the wicked (*tamas* results).

The qualitative attribute of the *guṇa*s is referred to as pleasure, stupor, or pain, but these affective characters are not mere emotional descriptions of states of mind. One can say that they are some sort of material causes of certain values or predispositional qualities called *bhāva* that are expressed when they are present. While Sāṃkhya texts do not directly elaborate the effects of the three *guṇa*s on different kinds of expressions in the phenomenal world, they insist that the presence of rajas in all possible combinations causes existence to be painful. The liberation theory of this philosophy rests on this pain thesis: Only freedom from all *guṇa*s can free us of pain.

3.6 Implications of the Idea of *Prakṛti* for Sustainability

Writers insist that the distinction of *prakṛti* in the texts of these philosophical systems is geared towards liberation theology (*mokṣa dharma*), and perhaps one should not stretch these concepts beyond their original interpretation. However, *mokṣa dharma* within some of these traditions has never been world denying. While Sāṃkhya philosophy clearly advocates liberation in some form, it also suggests a nuanced understanding of *prakṛti*. It is from this very understanding that one can derive moral or ethical frameworks.

According to Sāṃkhya philosophy, the material world is evolved from a preexisting ontologically real, material cause. This is called *sat-kārya vāda*. The effects exist potentially in the cause and thus are only transformations of the primordial materiality.[21]

The word *prakṛti* also carries a meaning that refers to Natural or *original*. In this sense, it is also an adjective-like. The adjective *prākṛt* derived from *prakṛt* means original or normal.[22] The meaning of *prakṛti* is "natural" in contrast to something that is an "imitation or a copy" called *anukṛti* or *kṛtrima* in Sanskrit (Apte 1986).[23] The name of the language that is derived similar to *prakṛti*—Prākṛt—usually refers to the natural language spoken by the common people. Historically, within a social context, Prākṛt was considered to be a degenerate form of Sanskrit by grammarians (see Deshpande 1993). The important idea is that it is not an "artificial" language that is contrasted with it, but a refined "well-constructed (*saṃ*)" language—Sanskrit, "*saṃs-kṛtam*".

---

[21]For a detailed discussion on tripartite process that links material effect and the material cause, see Larson and Bhattacharya (1987, pp. 99–101).

[22]*Prakṛta* is used in the geometry of altar construction where the measuring stick is divided into four lengths each called *prakṛta prakrama*—ordinary measure (from Jacobsen 2002, p. 31).

[23]And also from S. v "*kṛtrima*" in Monier-Williams, Leumann, and Cappeller, *A Sanskrit English Dictionary*, p. 303.

Thus, we find, in the light of Barnhart's (1997) views, it is true that for the Indian tradition "artificial" is in some sense related to the idea of reality rather than human contrivance. All human modifications of nature are only refinements and not re-creations. On the other hand, the idea of artificial is a reference to the fictitious, the non-true, and the imaginative. *Kṛtrima* would be "pretence" or an assumed reality. An adopted son for instance is referred to as *kṛtrimaputra* (Apte 1986).

The idea of "natural" and "artificial" in Indian thought may be of significance here. There is nothing that the human being can do to alter nature that will make it artificial and non-nature. This brings us to two conclusions. The first is that nature does not disappear but is modified and reformed/refined into other forms of reality. Restated in the language of *Sāṃkhya* philosophy, *prakṛti* becomes in parts *prakṛti* (the evolvable) *vikṛti* (the evolved, with no more possibility to evolve). The second conclusion is that given the close link between the idea of reality and of nature, the nature of reality in Indian thought especially with reference to *prakṛti* is fundamentally metaphysical. This insight into the interpretation of *prakṛti* has implications for the idea of conservation.

In the modern sense, there has been a sharp division between the categories of natural and artificial. We are clear that animals and other organisms are not like tables and cars. They are natural kinds. Lee (2005, p. 19) defines natural kinds and processes thus as "Naturally occurring entities and processes are precisely those which have come into existence, continue to exist, and go out of existence, entirely autonomously, and therefore independently of human intentionality and agency (and of supernatural agency for that matter)". An animal or a plant is also not like a lake, a rock, or a mountain that are also instances of natural kinds. On the other hand, artificial kinds would be those that exist dependent on human intentionality and agency.

This interpretation of natural and artificial is conceptually problematic in popular understandings of environment. There is a collapse between category nature–human and the category natural–artificial. While the first is a relationship between agency and object, the second is a category of modification; it is the natural that is processed into artificial. Without going into the details of the philosophy, just taking into account the idea that all materials are but modifications of *prakṛti*, it is simple to conclude that "clay" that we consider natural, and a pot that we consider "unnatural or artificial" are both *prakṛti*. There has been a change in the properties of the two, but intrinsically both are *prakṛti*. Such being the case, looking at the world through the Sāṃkhya perspective, there is a cause-and-effect relationship between natural and the artificial. If nature is all material on earth, the most artificial products of technology are also nature in some sense directly or as emergent. People for instance now see that there is "hidden water" in all products, such as the water (used to dye a pant) is hidden in the jeans pant or even a plastic button (as coolant for the mould).

The distinction between artificial and natural tends to separate the material cause from the material effect, giving more importance to the intelligent cause. In other words, people cannot see the cause in the effect or vice versa. To use a Sāṃkhya metaphor, it would be like refusing to see the mud in the pot while focusing on the potter.

The conversion of "natural" in the form of resources to the artificial is a category of modification; it is the natural that is processed into artificial. There has been naïve blindness in case of sustainability and conservation efforts. If nature is all material on earth, the most artificial products of technology are also nature in some sense directly or emergent. Again, quoting a Sāṃkhya example, milk becomes curds. The problem of diminishing resources is that the modification of *prakṛti* into artifical things is like milk being modified into curds.

This can lead to an argument that this would be dangerous to follow through in the context of ecological ethics. If everything was nature, there is nothing to conserve as such. But the next part of argument from the Sāṃkhya philosophy answers this objection. While it is true that it is *prakṛti* that is changed into all things we see around us, this is not a two-way reversible transformation nor is it a completely irreversible state of evolution. According to the Sāṃkhya philosophy, change is a formation of new collocations in the presence of concomitant conditions (*sahakāriśakti*) or efficient cause (*nimitta kārana*) (Dasgupta 1922). In normal conditions, these changes are bound by a law called *parināmakramaniyama* or the law of ordered transformation (unchangeable law). This is a limiting law based on the conditions of place (*deśāprabanda*) or limitation of time/season (*kāla prabanda*) limitation by form (*ākārā prabandha*) and finally limitation by causes (*nimittāpabandha*) (Dasgupta 1922). It is from the natural that we can create the artificial; the artificial is not the cause of the natural materially. Every object of human creation is created with the material cause of *prakṛti*. Or to put in the terms of modern ecology, the resources are *nimitta* (the cause) for all the objects around us, sharing a material relationship with the natural. This is ecologically relevant idea. Depletion of our resources—water, minerals, or forests—is bound by the limitation by causes rule. Our more than rapid rate of conversion of nature into its modified form is against conservation efforts. Seen from this angle, conservation is a slowing down of the change, not necessarily eliminating it. We could not probably turn the material artefacts back into their pristine original state (curds cannot become milk), but we could restore and rearrange the change again so we are not being the efficient cause of the rapid change to the natural around us. On the other hand, since nature evolves forming both resource (*prakṛti*) and a consumable or waste (*vikṛti*), one could reduce the second kind and look for resources to reuse, recycle. In fact, biodegradation is nothing but producing *prakṛti* evolutes from *prakṛti* as resource. What is also possible is using the laws of limitation to diminish the change or to cause only those changes that are not irreversible. Recycling water is one such example; actions like using alternate energy sources like solar energy would also come under this category.

3.7 Creation as the Body of God: *Srī-Vaiṣṇava* Tradition

Another important idea that is prevalent in Indian philosophical thought has to do with the whole creation as being the embodied form of god. This creates a common metaphysical grounding for nature reverence. Conceptually, the word *sarga* or *nisarga* for this school of philosophy captures this meaning of a "created–sacred universe". According to Mumme (2000), the *Srī-Vaiṣṇava* tradition describes the whole creation as emanation of the divine and the body of god. Though this seems somewhat similar to the idea of sacred nature, it is conceptually a metaphysical interpretation based on classical philosophical traditions and not based on just narratives and oral cosmology.

This idea is very prominent in the *Bhagavad Gītā*. There are verses that indicate that the world is pervaded by the presence of god. It is a form of panentheism where god is not merely the world, but as the origin and the resort of all beings in the world, he transcends them in not being totally embodied Mumme (2000, p. 138). The philosophy of theistic Vedanta (as opposed to the philosophy of *advaita*) especially the one propounded by Ramanuja is called qualified non-dualism—*viśiṣṭādvaita*. In this philosophy, the supreme *Brahman* is not just pure consciousness, but he is god with a divine personality—Viṣṇu—who has transformed into the world. This doctrine is called *brahmaparinamavāda*. *Viśiṣṭādvaita* philosophy explains the relationship between the cause and effect as different modifications of the same substance (Chari 2004). An often quoted example is that of the gold and its modifications as bracelet, ring, or chain.

The world is real, and it consists of the manifested part of *Brahman* (Sinha 2006, Vol. II, p. 653).[24] The relationship between the Brahman and the world is that of difference and non-difference—*bhedābheda*. At one level, the reality is a singular unity, and at the same time, it consists of multiplicity. Using the metaphor of the ocean and its waves, Ramanuja and his followers clarify that as the waves are identical to the ocean yet different from it, and the ocean cannot be limited to the waves alone, so also the Brahman and the manifest world are related to each other through identity and difference.

The distinction between conscious and unconscious parts of the universe is articulated by the concept of *cit*, the conscious, and *acit*, the unconscious. Matter which is produced by the transformation of the immutable does not have the subjective experience of I-ness (p. 673). The Brahman is associated with *cit* and *acit* in both their manifested form and the unmanifested form (Chari 2004, p. 272). The relationship between Brahman and the world is articulated as an embodied soul–body relationship (*śarīra–śarīri bhava*). The relationship is one where the like soul, as Brahman, controls and fulfils its own purpose through the body and the world. The Brahman is the basis and controller of the universe and is called *Iśvara*. The universe is made up of six metaphysical entities called *tattva*. *Tattva* is similar to the categories (*padārthas*). These *tattva*s are classified into substance (*dravya*) and non-substance (*adravya*). Technically, *dravya* is that which is substratum for modifications or states (*avasthā*s). Such modifications according this school of thought are explained as different from transitory relations that are not inherent and not the permanent nature (*dharma*) of a substance (Chari 2004). A pot on the ground and

[24]This philosophy provides an interesting alternative to the problems of *advaita* where the reality of the world is only an appearance. For more on *advaita*, also see Nelson (2000).

then same pot placed elsewhere is not *avasthās* of it. Nor is the cow-ness of a cow regarded as *avasthā* (p. 23). Chari (2004) adds "Only such an adventitious quality is inseparably related to the substance so long as the two last is regarded as an *avasthā*" (p. 23).

There are six *dravyas* according to this tradition divided into material (*jada*) and non-material substance (*ajada*)—God, individual self, knowledge, transcendental matter, cosmic matter, and time. The last two as *prakṛti*-cosmic matter and *kāla*-time are the material substratum of the universe. The nature of the universe—as constituted by *prakṛti*—follows the same metaphysics as that of Sāṃkhya philosophy. The concept of the world being the body of god has interesting implications for a theory of nature. The creation itself as originating in, sustained by, and identical to the supreme makes it sacred in a very substantive way. The human being like the world is a part of the same divine creation. The nature of the human being and the world is one such important theme that these philosophies explicate. Many of the philosophical schools may describe the various components of the creation that are ordered in a hierarchy of sorts. This order can be natural as in case of *Sāṃkhya* philosophy or it can be ordered by an intelligent being. The panentheistic interpretation of nature as a body of god has implications for issues around the place of the human and the non-human in the order of beings and is a rich source of eco-ethical insights.

References

Apte, V. S. (1986). *The student's sanskrit English dictionary*. 1899 reprint. Delhi: Motilal Banarsidass Publishers.

Barnhart, M. G. (1997). Ideas of nature in an Asian context. *Philosophy East and West, 47*(3), 417–432.

Bhattacharya, N. N. (1982). *History of the tantric religion*. New Delhi: Manohar Publications.

Bruun, O., & Kalland, A. (Eds.). (1995). Images of nature: An introduction. In *Asian perceptions of nature: A critical approach* (pp. 1–24). Richmond: Nordic Institute of Asia Studies, Curzon Press.

Chari, S. (2004/1988). *Fundamentals of Viśiṣṭādvaita Vedānta: A study based on Vedānta Deśika's Tattvamuktākalāpa*. New Delhi: Motilal Banarsidass Publishers.

Dasgupta, S. (1922). *A history of Indian philosophy* (Vol.1). 2004 reprint. New Delhi: Motilal Banarsidass Publishers Private Limited.

Deshpande, M. (1993). *Sanskrit and Prakrit: Sociolinguistic issues*. Delhi: Motilal Banarsidass.

Ghosh, J. (Ed.). (1977). Introduction. In A. Swami Hariharananda (commentary and translation) *'Sāṁkhya—sutras' of Panchashika and Sāṁkhyakhyatattvaloka*. New Delhi: Motilal Banarsidass.

Gottelieb, R. S. (Ed.). (2004). Introduction. In *This sacred Earth: Religion, nature, environment*. London: Routledge.

Halbfass, W. (1992). *On being and what there is: Classical Vaiaseosika and the history of Indian ontology*. Albany, NY: State University of New York Press.

Jacobsen, K. A. (2002). *Prakṛti in Sāṁkhya-yoga: Material principle, religious experience, ethical implications*. Reprint. New Delhi: Motilal Banarsidass Publishers Pvt. Limited, Indian edition. (First published by Peter Lang Publishing, 1999)

Kinsley, D. (1995). *Ecology and religion: Ecological spirituality in a cross cultural perspective*. New Jersey: Prentice hall.

Larson, G. J. (1979). *Classical Sāṁkhya: An interpretation of its history and meaning* (2nd ed.). New Delhi: Motilal Banarsidass.

Larson, G. J., & Bhattacharya, R. S. (Eds.). (1987). *Sāṁkhya: A dualist tradition in Indian philosophy, Encyclopedia of Indian philosophies* (Vol. 4). Delhi: Motilal Banarsidass Publishers.

Lee, K. (2005). *Zoos: A philosophical tour*. Hampshire: Palgrave Macmillan.

Mohanty, J. N. (1999). *Reason and tradition in Indian thought: An essay on the nature of Indian philosophical thinking*. Oxford, England: Clarendon Press.

Mohanty, J. N. (2000). Moral philosophy. In *Classical Indian philosophy*. Maryland: Rowman & Littlefield Publishers Inc.

Monier-Williams, M., Ernst, L., & Carl, C. (1990). *A Sanskrit English dictionary*. New Delhi: Motilal Banarsidass Publishers. (First Published by Oxford University Press, 1899)

Mumme, P. Y. (2000). The potential contribution of Śrivaiṣṇavism. In L. E. Nelson (Ed.), *Purifying the earthly body of God: Religion and ecology in Hindu India* (pp.133–163). New Delhi: D K Printworld (P) Ltd.

Narain, H. (1961). Finding an English equivalent for 'Guṇa'. *Philosophy East and West, 11*(1/2), 45–51.

Narayanan, V. (2001). Water, wood, and wisdom: Ecological perspectives from the Hindu traditions. *Daedalus—Journal of the American Academy of Arts and Science* (online). www.amacad.org/publications/fall2001/narayanan.aspx (Accessed 25 May, 2004).

Nelson, L. (2000). The dualism of non-dualism: Advaita Vedanta and the irrelevance of nature. In L. E. Nelson (Ed.), *Purifying the earthly body of god: Religion and ecology in Hindu India* (pp. 61–88). New Delhi: D.K Printworld (P) Ltd.

Potter, K. H. (1977). *The encyclopedia of Indian philosophies: Indian metaphysics and epistemology, Vol. 2 of encyclopedia of Indian philosophies*. New Delhi: Motilal Banarsidass Publishers.

Potter, K. H. & Bhattacharyya, S. (Eds). (1993). Metaphysics. In *Indian philosophical analysis: Nyāya-Vaiśeṣika from Gaṅgeśa to Raghunātha Śiromaṇi, Vol. 6 of encyclopedia of Indian philosophies*. New Delhi: Motilal Banarsidass Publishers.

Rao, K. B. R. (1963). The Guṇas of Prakṛti according to the Sāṃkhya Philosophy. *Philosophy East and West, 13*(1): 61–71.

Rolston, H. I. I. I. (1987). Can the east help the west to value nature? *Philosophy East and West, 37* (2), 172–190.

Sarukkai, S. (2002). *Translating the world: Science and language*. Lanham: University Press of America.

Sarukkai, S. (2010). Translation as method: Implications for history of science. In B. Lightman, L. Stewart, & G. McOuat, (Eds.), *Circulating knowledge* (to appear). Leiden: Brill Press. Also as preprint paper, *Translating concepts: Some issues in the methodology of history of non-western science*, presented in the Conference Circulating Knowledge, East and West, University of King's College Halifax, Nova Scotia, Canada, July 21–23 (By the courtesy of author).

Singh, R. P. (2005). Environment and vedic literature. In D. C. Srivastava (Ed.), *Readings in environmental ethics: Multidisciplinary perspectives* (pp. 106–114). Jaipur: Rawat Publications.

Sinha, J. (2006). *Indian philosophy* (Vols. 1 & 2). 1952 reprint. New Delhi: Motilal Banarsidass Publishers.

Sinha, N. (Trans.). (1979). *The Saṃkhya philosophy* (2nd ed.). 1915 reprint. New Delhi: Munshiram Manoharilal Publishers Pvt. Ltd.

Chapter 4
Nature as Elemental: The Matter of Nature

Abstract The idea of nature as constituted by five elements, or *pañcamahabhūtas*, is very popular in many naturalistic philosophies. Such a universe consists of both living and non-living parts of the cosmos. These elements are imagined as intangible to begin with, and then, they are understood as combining to form the gross elements which make up the cosmos. Traditions such as Vaiśeṣika and Saṁkhya explain the nature of these five fundamental elements and the process of creation of manifold diversities within them. This chapter will describe the materialism within Indian traditions and also dwell on the Vaiśeṣika atomism and the evolutionary nature of Saṁkhya tradition.

Keywords Elements · *Pañcamahabhūtas* · Matter · Creation · Earth · Water · Fire · Air · *Ākaśa* · Atoms · Transformation · Subtle elements · Gross elements

4.1 Pañcabhūtas

The word *bhūta* is used to denote "being" or "existence" and also refers to past tense in Sanskrit. The connotations of this root *bhū* imply creation, manifestation, and evolution. According to Filliozat (1996), there are multiple meanings of the word *bhūta*. He writes:

> In this rich semantic material four layers are prominent. The first centres around the idea of the past, the second around the idea of reality, the third around the idea of the living being, natural or supernatural and the last around the idea of inanimate thing and especially the element of matter (p. 53).

One can assume that since beings, like *bhūta*s (unlike the Supreme Being), are born, exist for a period of time, and will one day cease to be, they are "created existents". In Sanskrit, the word *bhūta* can also specifically refer to a class of spirit beings who are regarded as attendants of Lord Śiva.

M. Baindur, *Nature in Indian Philosophy and Cultural Traditions*,
Sophia Studies in Cross-cultural Philosophy of Traditions and Cultures 12,
DOI 10.1007/978-81-322-2358-0_4

For Halbfass (1992, p. 50), the terms *bhūta/mahābhūta* are said to "have an evolutionary connotation insofar as it suggests something that 'has become'". The other important meaning of *bhūta* is "material element". There are five fundamental elements, each of which is qualified as a *bhūta*. These elements are the fundamental types of material that make up the universe and therefore find mention in most classical Indian philosophies. Earlier Vedas such as the *Ṛg* (henceforth, as *RV*) refer to these elements but do not call them specifically as *pañchabhūtas*—a set of five elements. Also, not all schools of philosophy accept all the five elements. There are those like the materialists traditions for instance who do not accept space as an element.[1]

Lalye (1995, p. 65) establishes that the collective word appears only in later texts such as some of the Upaniṣads. We do however find many references to individual elements in Upaniṣads. Many of these references to elements are found in the themes of cosmogony, particularly questions of the origins of the universe. In the Vedas and Upaniṣads, there does not seem to be any specific description of the sequence of the origin of the elements, nor is there any enumeration of them as three or five. Later Upaniṣads, as the *Sāṃkhya Āranyaka* are clearer in their description probably after the rise of classical traditions (Lalye 1995).

Vedic texts often speak of one primordial being who is the source of both the intelligent and material cause of the creation. As the explanation of creation is based on a presupposition that it evolves out of one unitary being or materiality, the origin hypothesis must attend to a description of the process where diversity or the many is created from one. Halbfass (1992, p. 50) states:

> The individual elements are, of course, not necessarily presented as successive stages. From an early time on, they may be coordinated "horizontally" ***within*** "vertical" schemes of evolution. This is an example of "horizontal evolution" or "ramification" at one level of cosmic development. The elements evolve out of activities of the unitary supreme-being referred to as *Prajāpati* or as *Hiraṇyagarbha*.

Also required for such a world view is an explanation of the evolution of the gross (tangible) universe from a subtler being. Embedded in these theories are also hypotheses that explain the division of the conscious and the unconscious matter and distinguish the sentient from the insentient.

There are references to the individual elements in many of the Vedic and Upaniṣadic verses. The Vedas refer to these five elements, but not as a group or as the set of five elements. The five elements in the Vedas are not similar in their conceptualisations. The fire and wind elements are two of the primary deities of the Vedic pantheon. The Earth as a goddess is also accorded a status of a singular deity. The abstract sky and the waters have divine controllers but are not themselves clearly deified.

---

[1]The Cārvākas for instance believe in only four elements. They do not admit inference as a valid proof or meaning of knowing. Since space is intangible and cannot be grasped by senses, they do not admit that space can be an element.

4.2 Agni and Vāyu

4.2.1 Agni

Fire as an element in the Vedas is a very popular god with many hymns praising him. He shares a common origin with Indra being born of the mouth of the cosmic person … (*RV* X.91.13, trans. Griffith 1973). The connection between fire and water is often alluded to in some hymns. Logically, this seems an impossibility, but Staal (1996, p. 208) suggests that this is causal link. Since fire arises from wood, fire resides in plants. And since plants have water as their essence, the Vedic composers have made a connection with water and fire too. The same is true in the case of clouds, lightening, and fire. The deity of lightning, Mātariṣvan, brings down the heavenly fire to the Earth. All these origins have an empirical component to them showing that there was an attempt to connect elements based on the locus of origin, and not merely on the idea of a manifested quality. This might have been a precursor to the *tanmātra* concept of the Sāṃkhya philosophy.

The form of the deity *Agni* is not as anthropomorphic as gods like Indra. He is often called headless and footless and is considered in the form of flames:

> He is called butter face, butter backed, butterhaired, flamehaired.… His food is ghee, but he also devours woods and eats thrice a day.… *Agni's* Brightness is mentioned; he is like the sun, dispels darkness … but when he drives through the forest … his path is black. His flames soar terribly; his life reaches the sky (Keith 1925, p. 154).

The fire is supposed to have either seven flames or three flames, and the fire element is present wherever there is any heat. There are three kinds of fire classified functionally—the sacrificial ritual fire, the household fire and the fire of the funeral pyre. There are also different forms of the fire in the heavenly worlds (*śuci*), in the earthly spheres (*pavamāna*) and in the underworlds (*pāvaka*). The fire god is important in the Vedic world view for all their rituals as he is the messenger between the earthly plane and the heavens, who carries or conveys the offerings to the gods in heaven.

4.2.2 Vāyu

The wind element is a divinity in the Vedas. According to Keith (1925), the air element is referred to as both *vāyu* and *vāta*. While *Vāyu* is the more anthropomorphised deity, *vāta* is more elementary or natural. *Vāyu* is referred to as a form of the supreme *Brahman* that is presentational. Keith (1925, p. 139) describes the form of the air as *vāta:* "*Vāta* is merely the wind in its power, sweeping along great clouds of dust, shattering and thundering, his form cannot be seen by the mortal eye, though his roaring is heard, nor is the place of birth known". The wind is also linked to the rain god *Parjaṇya.* This is a very natural connection as clouds are carried along by the wind. The air in many Vedic hymns is also known as a messenger of gods and is invoked as medicinal healer:

Tvaṃ hi visvā bhesśajo devanaṃ tu dutāsi
Thou art the healer, the messenger of gods (*RV*, XI137.3, trans. Griffith 1973)

In Āyurveda, *vāta* becomes one of the three physiological fluctuations (*dośa*) responsible for many movements. Particularly, *vāta* is responsible for dividing the embryo into parts within the womb. An earlier hymn in the *Ṛg Veda* refers to this.

So stir thee baby unborn (RV V.78.7-8, trans. Griffith 1973)

4.2.3 Earth, Water, and Space

Gombrich (1996, p. 371) mentions that the Earth does not seem to have an anthropomorphic form in the Vedas. The *Śatapatabrāhmana* refers to the Earth as a *maṇdala* (circular). The early Vedic descriptions of the Earth are naturalistic and refer to it as carrying the hills and mountains. This is the idea of the Earth as supporting the land below our feet, as opposed to the sky above. We occupy the intermediate space. Earth, as *bhūmi*, in its forms includes rocks, stones dust, and the land (*Atharva Veda* XII.1.26). In the Yajurveda (*YV* XII.18), we find that there is a direct injunction not to hurt the Earth: "*Pṛtvīm mā hiṃsiḥī*" (do not do violence to the Earth). It is in the Purāṇas that we find reference to the Earth as goddess.[2]

The element water is often referred to in its plural feminine form as "waters". They are goddesses and flow into the sea. Waters are related to the ocean and other landforms as mothers or wives; it is very clear that the nourishing fertility of water makes it easy to imagine the waters thus. All waters are functionally meant to purify and cleanse. The use of water in rituals was prevalent, and even symbolic sanctification, consecrations and the everyday rituals were done by offering water. The waters not only cleanse the physical body but also purify the demerits, particularly if linked to a sacred source are sanctified by chants. The anthropomorphic forms of the waters are the *apsarās* (Keith 1925, p. 141). According to Oldenberg (1998, p. 126), they are water nymphs that move along the waters. Later interpretations give the sources of the waters—particularly the rivers—personalities and unique names. The deity Varuna is considered to be the controller of the waters in the Vedas. In the Ch. Up, it is said that the subtlest part of the food is consumed by the mind, while the subtlest part of the water is consumed by the vital force (*Ch. Up.* VI.7.6, trans. Ghambirananda,1983 p. 452). The link between water and life is found even in descriptions where water is called the essence of the Earth that makes up the beings.

---

[2]See Baindur (2010) for a discussion of the idea of the Earth as a sacred. A brief summary of this paper is also given in Chap. 8.

Ākaśa is known by many terms in the Vedas. It is still not clear whether the various meanings of this element and other terms that represent it can be subsumed under a general category of the element space. This is because there are so many meanings and contexts to the terms referring to the idea of space. Charkarabarti (1996, p. 109) suggests that the words for space are used in three layers of meaning that cover a whole range of interpretations—primary, symbolic, and metaphorical.

The earliest references to space in the Vedas are closer to the meaning of "sky" rather than space. The possible earliest reference is to the divine pair *dyāvāpṛthivī*, the Sky–Earth parents of the world. The intermediate space in between them is the first conceptualisation of space. The word for space may have come from idea of a gap or a hole, given by the term *kha* which stands for apertures or the axle hole of the wheel. The term *antarikṣa* is used to denote space in the Vedas, and the term *ākaśa* occurs in later texts. From a passage on meditation about space, we find that space is conceptualised as something that allows for freedom of movement (*Ch. Up.* I.9.1, trans. Gambirananda). We find reference to the term *ākaśamatra* in the *Prasnopaniśad* text, which refers to the subtle form of space as element, in contrast to mere *ākaśa* which is the gross element.

4.3 The Relationship of Elements in the Upaniṣads

Many verses in the Upaniṣads try and make interconnections between the diverse elements of the universe. This is an early attempt to claim unitary origins of the creation and a kind of evolution. In some Upaniṣads, for instance, the elements are included in a series of different aspects of the universe each of which forms the basis of another or its essence, or as each constituting a part of the supreme:

> Of these objects the earth is the essence; of the earth, water is the essence; of water, herbs are the essence; of the herbs the human body is the essence... (*Chāndogya Upanṣad* I.1.2, henceforth, *Ch. Up* trans. Ghambirananda, p. 8).

In verse VI.2.3, The *Chāndogya Upanṣad* states that as sweat is produced from heat in the body, water is born of the fire:

> That (Existence) saw, 'I shall become many. I shall be born.' That created fire. That fire saw, 'I shall become many. I shall be born.' That created water ... water comes from heat (*Ch. Up.* trans. Ghambirananda, p. 422).

In the same chapter, the idea that all of the elements have a unitary subtle composition of the elements is indicated using the example of the fire:

> The red colour that (gross) fire has, that is the colour of (subtle) fire. That which is the white colour (of the gross fire), that is of (subtle) water. That which is the black colour (of the gross fire), that is the colour of the (subtle) earth... (*Ch. Up.* VI.3.4, trans. Ghambirananda, p. 434).

In the *Taittarīya Upanṣad*, there are many meditations describing the "great combinations". Among these is a reference to the combination of elements. In each of these combinations, there is one prior form and then a later form that form a link or junction referred to as *saṇdhi*.

One of these verses reads:

> As regards the meditation on the worlds: the earth is the first letter. Heaven is the last letter. The sky is the meeting place. *Vāyu* is the link (*Ta.Up.* I.3.2-4, trans. Ghambirananda, p. 434).

It is important to understand why the *Upaniṣad* uses the term *saṇdhi*. *Saṇdhi* refers to the phonetic combination of precedent and antecedent letters in two sequential words in Sanskrit language. For instance, if the precedent word is *hima*, the antecedent is *ālaya*. Their *saṇdhi* product is the word *himālaya*, the point of contact being the long vowel "*ā*" sound. One could posit that the combination of the elements is analogous to the combination of the elements. The implication of this is that there has to be a series of elements with a prior and a later element and the product has the qualities of both with no suggested or emergent meaning (as it is in the case of compound words formations called *samāsa*, which are meaning-dependent). The combination product conjoins the properties of both the combinants, the changes occurring only at the point of contact, (called "link" in the translation of the verse above) the *saṇdhanaṃ*. In the above verse, the contact itself is referred to as Saṇdhi Contact and the point of contact is the *saṇdhanaṃ*. Another meaning of *saṇdhanaṃ* is "forge" or the locus of a joint. From this understanding, we have an interesting imagery of the elemental combinations. The prior is the Earth and the latter is the heaven. The contact is the space, and the locus of the joint or union is the wind. Similarly, in another verse, the combinants are the fire and sun, the combination is the water, and the locus of the joint is the lightning. In these descriptions of the elements and the other combinations, the *Upaniṣad* describes instances of other combinations including the elements, loosely classifying them as belonging to the different realms or spheres of experience such as the physical, the luminous, or the experience of learning or birth of a child. There seems to be an attempt to deconstruct the universe to fundamental units of creation that form a unitary whole.

What is interesting in this metaphorical image is that the process of deconstruction here is analogous to language or the human body both of which have joints, or *saṇdhi*. The *Upaniṣad* does not use the metaphor of a pot or any other mechanistic image to describe these combinations. The idea of the body image seems relevant here because there are already references to the creation as a cosmic person, sometimes referred to as *Virāt*, or *Hiraṇyagarbha*. The elements then become the various bones of the supreme, the joints being the *saṇdhi* and the exact location of this joint is the *saṇdhanaṃ*.

Other verses in many Upaniṣads describe the formation of the elements. These preliminary conclusions about the formation of a universe that is available to our senses influence later classical philosophies that develop these ideas more concretely. Particularly, the creation of the five elements mentioned in the *Aitareya Upaniṣad*, verse I.i.4, seems to be a precursor to the Sāṃkhya theory about the creation of elements. This particular verse describes that the elements arise from the body of the Virāt of the human form, speech is born from the mouth and from that speech, the element fire is born. Similarly, the division of the nostrils creates the sense of smell and from the sense of smell, comes *vāyu*. Likewise, the partition of the ears produces sense of hearing that gives rise to *dik*, directional space. The passage goes on to describe the production of all the organs including the mind from the heart and finally procreative organ from which water emerges.

In the Upaniṣads, we find the individual references to each of the elements and there are many different narratives of their origin. Here, the elements are not referred to as deities but as natural elements. *Muṇḍaka Upaniṣad* has a verse that everything originates from the *Puruṣa*, the Supreme Being:

> From Him originates the vital force as well as the mind, all the senses, space, air, fire, water, and the earth that supports everything (*Mu. Up.* II.1.3, trans. Ghambhirananda 1957a, p. 112).

4.4 The Natural Elements in Sāṃkhya

One of the most sophisticated descriptions of the five elements and their evolution is found in Sāṃkhya-Yoga philosophy. The process of creation is very clearly explained in the Sāṃkhya texts which postulate these elements as *tanmātras*, or subtle elements, and five gross elements as evolving out of the primary cause *mahat*. *Mahat* itself is the evolute of *prakṛti*, the primordial nature.[3]

We have already seen how the idea of *prakṛti* in Sāṃkhya is described as made up of three co-constituting *guṇa*s and also how the entire created world evolves from the primordial reality called *Mūlaprakṛti*. In this section, we shall particularly see how Sāṃkhya philosophers describe the evolution of the five elements and the creation of the coarse/tangible universe from the primordial nature. In the *Yuktidīpikā* (henceforth, *YD*), one finds that there are particular beings whose bodies are directly created from the elements. Other beings arise out of the body of a mother and a father, but these beings are called *prabhūta*. These element-born beings are either seed-born or sweat-born.

Sāṃkhya particularly distinguishes between subtle and gross elements. While the tangible world is made of the coarse or gross elements, the subtle elements are causal in creating the gross elements. Aniruddha, the commentator, in his *Sāṃkhya Sūtravṛtti*, points out that the manifest world is made up of the gross elements (Larson and Bhattacharya 1987, p. 349). The creation of the gross elements from the subtle elements and their relationship to the senses is a recurring theme in many philosophies. The purpose of the created elements is to function for the sake of the consciousness, the *Puruṣa* (cited in Larson and Bhattacharya 1987, p. 346):

---

[3]This has been discussed as one of the main conceptualisations of nature in Chap. 3.

> The effects of creative materiality are inclusive of the intellect through the five gross elements.... All of these effects function for the sake of the consciousness, but only as mediated through the activity of materiality.

4.5 Creation of the Elements from Primordial Materiality

Sāṃkhya philosophy explains the many evolutes of *prakṛti* as a complex set of *tattvas*, or principle entities: "The origin of all *Tattvas* is *Prakṛti*, the primordial nature" (*Sāṃkhya Kārika* 3, henceforth referred to as *SK*, XXII p. 22).

Larson (1979, p. 178) suggests that the analysis of the world in Sāṃkhya is from the perspective of the *puruṣa* who is both individual and impersonal: "In other words, world is comprehended in terms of how the *puruṣa* witnesses it. This explains why the principles (*tattva*s) in the *kārikā* are expressed in terms of the psychological rather than cosmological".

This view according to him is not particularly an experimental or scientific research of the psyche, nor is it an explanation of what the *puruṣa* experiences (as *puruṣa* actually does not act, being the witness principle).

Mohanty (1999, p. 209) points out that the creation of elements starts with the constituents of *prakṛti*, the three *guṇas*:

> On the Indian theory, the root Nature, the unmanifested ***prakṛti***, does not consist of atoms, only not yet forming heterogeneous combinations. It rather consists of what are called *guṇas* (provisionally to be rendered qualities). These ***guṇas*** are said to be three: ***sattva***, ***rajas***, and ***tamas***. These constituents, we are told, are of the nature of pleasure, pain, and indifference respectively.

The manifest world arises from the unmanifest, and the first of these *tattva*s (translated as principle or entity) to arise from *prakṛti* in the presence of the *puruṣa* is the intellect principle (*buddhi*). This is not the thinking faculty of the mind, but the *buddhitattva* posited as the intelligence principle of creation itself, the guiding rationale that determines the ordered progress of creation. Larson (1979, p. 89) points out that like the *puruṣa*, "... the *buddhi* is individual yet impersonal". One must also note that the *buddhi* is *jada*, unconscious except in the presence of *puruṣa* that (is uninvolved in action, but) supports the evolution of *prakṛti*. This is also called *mahat*, or great. The *mahat* precedes the egoity or awareness of the self (*ahaṃkāra*) and so it overrides the idea that the world is a personal creation. The *mahat* can be conceptualised as a master plan in potential, ready for the egoity to manifest itself. Larson (1979, p. 180) suggests that this might be an idea that is derived from earlier cosmological accounts of a creative principle.

The process of creation is intelligent in the sense that it is full of resources (*aiśvarya*), knowledge (*jñana*), ascertainment (*adhyavasāya*), and order (*dharma*). This is the sattvic form of *buddhi*, the tamasic form being the opposite. It is not clear from texts whether these are constituents of *buddhi* or qualities. Given the Sāṃkhya understanding of *prakṛti*, it is likely that these are transformative and co-

constituted by the *guṇas*. Having these features does not imply will or any form of personal intelligent activity we associate with mind functions.

Another significant point to note is that Sāṃkhya philosophy does not have a divine god or overseer. In some sense a mechanistic intelligence that functions in the presence of a conscious witness entity that accounts for the activity of creation. Under such a conceptualisation a personhood of a creator seems to be unnecessary. It is easy to see how the principle of the *mahat* can easily be equated to that of the Supreme Being or God in later Yoga philosophy.

From *mahat* arises egoity (*ahaṃkāra*), a principle of individuation. Larson (1979, p. 186) points out that this egoity is not a personal sense of the self but a self-awareness that pervades all of the experiences including functions of the mind and senses (p. 186). From *ahaṃkāra,* a twofold creation takes place. The first consisting of the "group of eleven" arises from the sattvic *ahaṃkāra* consisting of the five action-senses (*karmendriyas*), the five intelligent senses (*Jñanendriyas*), and the mind (*manas*). From the tamasic form of egoity (*ahaṃkāra*), the five subtle entities, the pre-elements called *tanmātras*, arise. The *Yuktidīpika* equates the *tanmātra*s to the *pañcabhūta*s, but most other commentaries emphasise that the *tanmātra*s are not elements but potentials that could evolve into the elements.

The idea of *tanmātra*s is particular to Sāṃkhya. On one hand, this principle may have been introduced to evidence *satkāryavāda* doctrine by insisting that the elements pre-exist in the egoity as potential objects for the senses. The concept of *tanmātra* creates the link between the senses and the five elements in the Sāṃkhya school.

The mind and the 10 senses of grasping, or *indriya*s, constitute the instruments through which an embodied *puruṣa* experiences the external world. The senses created from *ahamkāra* at this point of evolution are not the end organs or embodied senses, but they are senses in their potentiality. Similarly, the *tanmātra*s are also intangible potentials. *Tanmātra* can be translated as "only so much" (Larson 1979, p. 187), or "measure of that much". Dasgupta (1987 p. 251) translates this as "potentialities".

Dasgupta (1987) points out that the derivation of the *tanmātra*s from *ahaṃkāra* has been considered by many Western scholars (p. 85) as the derivation of matter from ideas and thoughts. He argues for the materialism of Sāṃkhya evolutes by detailing the evolution of *prakṛti* into the five elements. According to him, each stage of evolution is not the creation of any new evolute but all are transformations of primordial *prakṛti* (pp. 85–86).

There are three stages of the evolution of the *tanmātra*s. The evolution occurs in stages that are called *bhūtadi*, *tanmātra*s, and *paramaṇu*s (pre-elements, subtle elements, and ultimate atoms. He writes that "the *bhūtadi*s are absolutely homogeneous with no qualities other than quantum…" (p. 86).

Seal (quoted in Dasgupta 1987, p. 87) suggests that *tanmātra*s are "infra atomic particles that are charged with specific potential energies". He understands the creation of each of these elements as a process in which each kind of *tanmātra* is charged with a particular form of energy that under suitable conditions form the respective atoms (*paramāṇu*). For example, the class of *tanmātra,* that is charged with vibration energy, forms the ether or space atom. Similar is the case with the classes of *tanmātras* that are charged with other energies that form the atoms of earth, water, fire, and air.

One can creatively suggest that the potential of "just-that-much" sound leads to the formation of its locus, the *akāśa*. And the potential of "just-that-much" smell leads to the formation of its locus, the air.

The *Yogasūtra* also mentions the elements as having different aspects during the description of focused meditative practices. Based on the idea of *Satkāryavāda*, the *Yogasūtra* advocates that the meditation on the past, present, and the future forms of the element would lead to the development of victory over elements. The *sūtra* iii.43 says: "Victory over the elements is achieved by *Saṃyama* (concentration) on the gross, essence and subtle form and the recurrence of *Guṇas*" (trans. author, from *Yogasūtra* 1976).

The gross form of the element is one of the aspects in which the elements are found in combination with each other. The second aspect of the element is the special or its essential property, such as heat in the fire. The third aspect of the element is the *taṇmātra,* the material cause of the element during evolution. The *Yogasūtra* commentaries claim that the *taṇmātra* is a kind of generality that manifests itself as many particularities called *viśeṣa.* For example, if sound is the *taṇmātra* for the space, the different notes are particularities.

Sāṃkhya interprets the entire universe as being geared to be available to the apparent experience of the *puruṣa.* The sensations and their potentials lead to the formation of a tangible world. Again as Dasgupta (1987) in his study has affirmed, this is not a theory of psychological creation where the material arises from the immaterial. Rather in this step of the transformation of *prakṛti*, the link between the outer world of experience and the inner world of senses is established. The simultaneity of the sense and the sense object potentials creates a common ground that is important for an explanation of how the senses function efficiently. Each sense seems to lock on to its own object and is unable to grasp other objects of other senses. What forges the link between these senses and their objects? According to the *Satkāryavāda* doctrine, while the link between the different objects and the experience as a whole is ascribed to the mind, it seems that for the Sāṃkhya philosophers, the sense—sense object fit is possible only if they both have arisen from the same kind of material.

From a particular perspective, without subscribing to Cartesian dichotomies (that are absent in Sāṃkhya), one might understand the evolution of the potential *tanmātras* and the disembodied sense potentials as a kind of divisive conceptualisation of the subjective and objective entities. The difference between *sattva* and *tamas* predominance is used to separate the instruments of cognition—the group of eleven senses—from the *bhūtadi*, the precursors to the material created world. Both of these are evolutes of *prakṛti* differing only the aspect that gives rise to them while the *sattvika ahaṃkara* creates the grasper or experiencer, the *tāmasika ahaṃkara* transforms into the experienced nature or the objective world of reality.

From the Sāṃkhya view point, both these evolutes of *prakṛti* are *jada* or unconscious (thereby they are not strictly subjective and objective divisions of the world). The sentient *puruṣa* is also not a subject but is made a reluctant subject due to the activities of *prakṛti*. *Prakṛti*, on the other hand, is unconscious and is made active in the presence of the *puruṣa*. A metaphor of reflective capacity of the different evolutes would help clarify the position of the group of eleven and the *bhūtadi*. One could say that the evolutes of the *sāttvika ahaṃkāra* reflect the sentience of the *puruṣa* better, while the tamasic predominance makes the *bhūtadi* almost opaque, and hence, they occur as insentient (*jada*). True nature as *prakṛti* is essentially *jada* and all the sentience we find is a reflection of the *puruṣas*. This argument falls into the general theory of reflected *puruṣa* in the sattvic predominant *citta*, a much discussed topic in the theory of *Yoga*. The implications of this for the conceptualisation of nature are very interesting. Nature as material and insentient is its essential form, and the sentient part of beings, including the human beings, is actually "non-nature" as they in their embodied states are reflections of the *puruṣa*.

Many of the earlier Sāṃkhya texts just mention that the five *mahabhūta* give rise to the gross elements called the *pañcabhūtas*. This transformation of the *tanmātras* into the material elements takes place through a particular process often called in later texts as *pañcīkaraṇa*. Later Vedānta philosophies adopt these explanations into their doctrines.

Each element in the subtle form as *mahābhūta* first divides itself into two equal halves. The element retains one half of itself, and the other half is distributed to the other four elements in equal portions. Each gross element therefore is constituted of one half of its *mahābhūta* form, and one eighths of the other four elements. The gross elements make up the known objects of the world.

4.6 Elements in Nyāya and Vaiśeṣika Schools of Thought

Though Nyāya and Vaiśeṣika schools are syncretised in many respects, when it comes to the theory of change, they have two different perspectives. Therefore for clarity, in this particular section, we shall treat them as two individual and different schools of thought.

The elements in both Nyāya and Vaiśeṣika philosophies are necessarily connected with a sense faculty. The Nyāya school defines *bhūta* as that "which has a specific property that can be grasped by an external sense organ". As we have seen earlier, things or *dravyas* are made of atoms of different elements. The elements (*bhūtas*) are described as five in the *Nyāya Sūtra*, 1.1.13: "The material substances are Earth, water, fire, air and space" (p. 2430, trans. Jhā 1950).

The classification of the elements in the Kaṇāda's *sūtras* (Vaiśeṣika school) begins with a description of their properties summarised from *sūtras* II. 49–53 (trans. Chakrabarty 2003, p. 49):

> There are no qualities in the *Ākāśa* (space), *Vāyu* (air) is touchable, while fire has both *rūpa* (colour/form) and touch. Water has *rūpa* (colour/form), touch, taste and fluidity and smoothness. The element that has all the qualities of water and also odour is earth.

Kaṇāda points out that the specific property of each element is its mark (*Liṇga*). The mark in Indian philosophy is something that distinguishes one object from another. Hence, the distinctive mark of the five elements is the presence of these properties. Since the properties are related to particular senses, it would not be wrong to say that the world is made of objects that are effable and knowable, particularly through the senses. Kaṇāda says in *sutra* ll. 1. 8–7 (Chakrabarty 2003, p. 50):

> Horns, hump, hair at the tip of its tail and a dew lap are the visible signs of a cow. (Similarly) touch (is the sign) of *Vāyu* too. And this touch is not of the visible (things) hence *Vāyu* has an invisible mark.

Sharma (2000, p. 177) summarises:

> The peculiar qualities of earth, water, fire, air and ether are smell, taste, colour, touch and sound respectively which are sensed by the five external senses. The external senses are constituted by the respective elements whose specific qualities are sensed by them—the sense of smell is constituted by the element of earth and so on. The elements are the substrata of these qualities.

The various elements are said to be the material causes of the sense organs. For example, the visual organ is made of light (fireFire) and the tactual organ is made up of air (Sinha 2006, p. 424). Each element has one specific property that can be grasped by the senses. Since these properties are deeply related to the particular object of the senses, one could suggest that for the Nyāya and Vaiśeṣika philosophers, the experienced world is a sense-grasped world, where the senses distinguish between the different elements. The senses would also help us know the elemental composition of a particular *dravya*. For instance, anything with smell would necessarily contain the earth element and those with taste would contain the water element and so on.

Both Nyāya and Vaiśeṣika schools posit that *dravya*s can be made up of the combination of different elements because we can perceive the same object with more than one sense faculty. This is why we can sometimes "see" air that is coloured or "smell" water. For *Sāṃkhya* philosophy, the transformation of subtle to gross elements through *pancīkaraṇa* creates a sensible world. For the NV school, the composite combination with a predominant element creates sufficient magnitude of the atoms that are available to sense perception.

The properties of fire are important for the Nyāya school. How does the cooking of atoms produce a new substance? Fire becomes an element that has the capacity to change and substitute new qualities. Fire is also classified into four types based on its qualities (Sinha 2006, p. 384):

1. Fire manifested with touch and colour as properties, such as the sun and the flame.
2. Fire that cannot be seen but only available to the touch sense, such as in hot liquids and objects.
3. Fire with colour, visible but with no manifested touch. This is like the light from a lamp, (though the actual flame is of the first kind). Sunlight is also not this kind as it is warm and can be felt by touch.

4. The sense organ of the NV school is a special case as it is made up of the fire element but is not available to either sight or touch. The peculiarity of air and space is that they do not inhere in other *dravyas*.

The Vaiśeṣika account of elements suggests that the elements in their most fundamental form are *paramāṇu* or eternal atoms. These atoms combine to form composite wholes that are more than the sum of the parts. Both the Nyāya and Vaiśeṣika schools ascribe eternality to the atoms of the Earth, water, fire, and air. The composite forms of these elements are not eternal. Nyāya and Vaiśeṣika admit the reality of space as an external element (Sinha 2006, p. 402). Space is a special element that is eternal. For the Vaiśeṣika school, air and space are eternal because they do not inhere in other *dravyas* (*Kaṇāda Sūtras*, II.1.61, II.1.76). As we noted in an earlier chapter, a *dravya* whole can inhere in other *dravya* parts. Since air and space are partless, they do not have parts nor can they form composites. Where ever we find air or space, other elements are not present. They are in "contact" (*sannikarṣa*) with other substances and also as in the case of multiple forms of air, with their own kind (*Kaṇāda Sūtras*, II.1.62–63). Each of the atoms of the elements is possessed of the respective *jāti* (genus). For instance, the genus earth inheres in the earth element and so on. Earth atoms are responsible for formation of different types of bodies, including the human body. Nyāya and Vaiśeṣika schools posit that there are two kinds of bodies, those born of sexual union and those that are born of elements in combination with merits and demerits. Among the former, some bodies are born of a womb, others of eggs. Divine bodies and minute bodies are the latter kind (Sinha 2006, p. 383).

All tangible solid objects such as earth, stone, and plants that can be touched and also have odour are included under the earth category. In some texts, there are beings described as having elemental bodies, in combination with earth such as water-bodied and fire-bodied beings. There seems to be a similar idea in Jainism. As mentioned earlier, each one of the elements has a special property that is perceived by the respective sense organs.

4.7 Atomic Theory and Theory of Change: *Pākavāda*

The Nyāya and Vaiśeṣika schools consider the small, uncreated eternal substances as *aṇu* or atoms. Earlier Vaiśeṣika schools believed that all perceptible entities could be destroyed because they were formed by composites. Nyāya and Vaiśeṣika schools differ on the theory of composite formation.

According to the Vaiśeṣika school, atoms are too small to be seen within the threshold of perception. The *Kaṇāda Sūtras* (VII.1.8–9, p. 89) explain the two orders of magnitude and existence: *aṇutva* (atomicity) and *mahattva* (greater magnitude). The *paramāṇu* of atomic magnitude are eternal and imperceptible to the senses; the essential qualities of these elements atoms are also eternal, such as "earthness" or "waterness". The qualities of the created composite *dravyas* are not eternal, because the *dravya*s themselves are non-eternal (*Kaṇāda Sūtra*s, VII.1.4–5, p. 88). How does the change in these composites occur?

The *sūtras* seem to have a strong empirical foundation. They next describe that some earth-like materials such as beeswax and lac become fluid when in contact with fire. The importance of things being transformed by fire is explained as a way in which newer composites of materials are created. For this, Nyāya and Vaiśeṣika schools have different explanations. Anything that is heated by fire is called *pāka* or cooked. Kaṇāda says (in *sūtra* VII.1.6, p. 89)

> *Kāraṇguṇapūrvakāh pṛthiviyāṃ pākajāḥ*
> The cause of the earlier qualities are reproduced from the cooking in the earth (trans. by author).

The illustration is that of an unbaked black clay pot being "cooked" by the fire and changing all its properties and becoming red, baked, and capable of holding water. How do the transformations of the properties take place? The Vaiśeṣika school hypothesis is that the change occurs at the level of the *aṇu*, the atoms constituting the pot. Fire (by cooking) destroys the transient or non-eternal properties of atoms (such as the black colour), the eternal substance or the *dravya* atoms with their eternal properties continue to persist. When the heat produces a new set of properties in each atom, such as the red colour, they recombine to form the pot with the red colour and a new set of non-eternal properties (Potter, p. 84). This is known as *pilupākavāda* the (cooking of atoms) of the Vaiśeṣika school. In contrast, the Nyāya school proposes the *pitharapākavāda* (cooking of pot) hypothesis. According to this theory, it is not necessary for the changes to occur at the atomic level. The pot itself can undergo change and composites can also gain new properties.

This chapter has focused on two main themes through the discussion on elements. Firstly, after the basic description of the origin of elements, change and evolution of these elements and their relationship to each other in nature have been described in Indian philosophy. Secondly, the two theories of change demonstrate the concept of change in nature. The idea of fundamental elements that constitute nature creates the "few to many" problem. How do basic constituents contribute to the wide variety of materials in the world? And is there an intelligent cause required for this process? How does the transformation of substances occur in nature? These are some of the observations from some philosophical schools that were discussed in this chapter.

References

Baindur, M. (2010). Humanising the earth: reversing the 'reverence to resource' in the Indian context, Special Issue: Globalism, transnational, gender and ecological engagements. *Man In India*, 90(4), 567–587 (Guest Editor: P. Bilimoria, with assistance of Amy Rayner).

Charkarabarti, S. C. (1996). *Ākāsa*. In B. Bāumer (Ed.), *Kalātattvakośa*: *A lexicon of fundamental concepts of the Indian arts, Vol III, Primal elements-Mahābhūta*, K. Vatsyan, (Series Ed.).

New Delhi: Indira Gandhi Centre for the Arts, in association with Motilal Banarasidass Publishers.

Chakrabarty, D. (2003). *Vaiśeṣika-sutra of kaṇada* (translated with reference and index). New Delhi: D.K. Printworld (P) Ltd.

Dasgupta, S. (1987). *Natural science of the ancient Hindus*. New Delhi: Indian Council for Philosophical Research.

Filliozatt, P. S. (1996). Bhūta-Mahābhūta. In B. Bāumer (Ed.) *Kalātattvakośa: A lexicon of fundamental concepts of the Indian arts, Vol III, Primal elements-Mahābhūta*, K. Vatsyayan (Series Ed.). New Delhi: Indira Gandhi Centre for the Arts, in association with Motilal Banarasidass Publishers.

Ghambirananda S. (1957a). Taittarīya—Upaniṣad. In *Eight Upaniṣads, with the commentary of Śankarācārya*. (Vol. I). Calcutta: Advaita Press.

Gombrich, G. S. (1996). Pṛithivī/Bhūmi. In: B. Bāumer (Ed.) *Kalātattvakośa: A lexicon of fundamental concepts of the Indian arts,Vol III, Primal elements-Mahābhūta,* K. Vatsyayan, (Series Ed.) New Delhi: Indira Gandhi Centre for the Arts, in association with Motilal Banarasidass Publishers.

Griffith, R. T. H. (1973). *The hymns of Rig Veda*. New Delhi: Motilal Banarasidass Publishers.

Halbfass, W. (1992). *On being and what there is: Classical Vaiaseosika and the history of Indian ontology*. Albany, NY: State University of New York Press.

Jhā, G. (1950). Gautama's Nyāya Sūtras by Gautama, with the Bhāysa of Vātysāyana and Vārtika of Uddyothakara (trans.). New Delhi: Motilal Banarsidass Publishers.

Keith, A. B. (1925). *The religion and philosophy of the Veda and Upanishads: Part I*. New Delhi: Motilal Banarsidass Publishers.

Lalye, P. G. (1995). The panca mahabhutas. In J. V. Narlikar (Ed.) *The nature of matter* (Vol. 4, pp. 97–114), K. Vatsyayan *Prakṛti: The integral vision* (Series Ed.). New Delhi: Indira Gandhi Centre for the Arts.

Larson, G. J. (1979). *Classical Sāṁkhya: An interpretation of its history and meaning* (2nd ed.). New Delhi: Motilal Banarsidass Publishers.

Larson, G. J., & Bhattacharya, R. S. (Eds) (1987). *Sāṁkhya: A dualist tradition in Indian philosophy*, Encyclopedia of Indian Philosophies, (Vol. 4). Delhi: Motilal Banarsidass Publishers.

Mohanty, J. N (1999). *Reason and tradition in Indian thought: An essay on the nature of Indian philosophical thinking*. Oxford, England: Clarendon Press. Available from: http://www.questia.com/read/9827243.

Oldenberg, H. (1998). *The religion of the Veda* (S. B. Shotri: trans.) New Delhi: Motilal Banarasidass Publishers.

Sharma, C. (2000/1987). A critical survey of Indian philosophy (Reprint in 2000). New Delhi: Motilal Banarsidass Publishers.

Sinha, J. (2006/1952). *Indian philosophy*. (Vols. 1 and 2. Reprint), New Delhi: Motilal Banarsidass Publishers.

Staal, F. (1996). *Ritual and mantras: Rules without meaning*. New Delhi: Motilal Banarsidass Publishers.

Chapter 5
Nature in Some Philosophical Traditions of India: Basis for a Common Ground

Abstract Following on the same theme of Chap. 4, I address the ecological understandings of nature in the theories of Sāṁkhya philosophy and Advaita philosophy of non-dualism. This chapter also briefly examines other ideas of oneness of nature and human beings in classical schools of philosophy including Buddhism and Jainism. The concept of oneness given by the Sāṁkhya philosophical concepts of *prakṛti* and *puruṣa* is described. In this chapter, I have also suggested the moral implications of the co-constituents of *prakṛti*, called *guṇa*s. I suggest that the psycho-physical nature of *guṇa*s allows for new paradigms of environmental ethics. The idea of substantive oneness of creation based on *Advaita Brahman* is also discussed. Implications of such conceptualisations conclude this chapter.

Keywords Human–non-human · *Puruṣarthas* · *Karma* · *Guṇa*s · Substantive oneness · *Kṣetrajña*

5.1 Introduction

The multiplicity of belief systems in Indian traditions makes it possible to further explore in detail different perspectives that are not totally dependent on the concept of "nature" as informed by Western discourse.[1] As discussed earlier, categories and concepts of the cosmos or our habitat can be analysed at the level of creation, cosmos, earth, or even metaphysical mythical worlds called *loka*s. The way we conceptualise our habitat in traditional philosophies is not dependent on any particular discourse of nature that is exactly parallel to the Western concepts of nature. The interesting fact about Indian traditions of thought is that while a large portion of the philosophy deals with metaphysical concepts and epistemological

[1]In this case, I use "Western" to refer to the particular history of ideas I have traced in Chap. 2. As in case of the diversity of views in the world, there is no one homogeneous West or East.

M. Baindur, *Nature in Indian Philosophy and Cultural Traditions*,
Sophia Studies in Cross-cultural Philosophy of Traditions and Cultures 12,
DOI 10.1007/978-81-322-2358-0_5

issues, we also find the prescription of pragmatic practices that embody these concepts in norms articulated as ethical actions or restraints.

A critique of such a conceptual project is possible and may add to the environmental problematic rather than resolving it. The important question would be about the time and the contexts during which such concepts were evolved. In other words, one might ask, are such conceptual resources from ancient texts valid in the modern world? On the one hand, among current scholars, there has been much more focus on the problematic of such intersections and a critique of the positive implications of these belief systems. On the other hand, there has been a romanticised reading of the positive features of Eastern systems of thought that are portrayed as pro-ecological.

To use a medical metaphor, it is possible to re-examine some ancient herbal remedy for its ability to cure an ailment diagnosed modern times. We may, however, have to ignore its original prescriptive functions in the light of modern understanding of plants. Similarly, it is possible to look for conceptual resources that underlie the cultural ethos of a community and seek to understand the relationship between humans and nature, to find solutions for modern ecological problems. Such concepts may have to be reoriented ecologically and analysed. It is also widely acknowledged that many presuppositions, unacknowledged, and undistinguished inform many ethical actions and behaviour in India. Even to distinguish and recognise these beliefs may help to organise conservation movements or educational awareness efforts in a better way. Rolston (1987) remarks:

> So it might be that an Eastern model of nature can critique the metaphysical assumptions in evolutionary ecoscience and in technological science and thereby help the West to value nature. Perhaps an Eastern metaphysics can offer a better model to go with Western science, interpreting that science and making it ethically responsible before its environment. But we cannot presume any such insight (p. 174).

The important thing to note here is that *conceptual resources* are *conceptual* in the sense that they are only ideas. The understandings of concepts are not geared to direct practical uses, but they inform practices, help us understand the problems from a different perspective, and look for solutions that are unusual. It is important to stress here that conceptual resources are not solutions to pragmatic problems.

Roach (1996, p. 53) suggests that various conceptualisations can be modified in such a way to turn them into ecologically relevant ideas for current times. She uses the metaphor of bio-degradation to suggest that ideas can be "biodegraded" so the resultant discourses are environment friendly. So my attempt here is to biodegrade some of these conceptualisations from Indian philosophies and traditions of thought, and check if the resultant alternative ideas are applicable to some of issues of nature–human relationship discussed in Chaps. 1 and 2.

Philosophical ideas around nature are usually placed within some environmental ethical positions describing a theory in terms such as "bio-centrism", "ethnocentrism", "anthropocentrism". These are newer ways of understanding the earlier ideas about nature. My overview ignores these positions (which are sometimes political) and attends to philosophical perspectives and at the same time tries to place the conceptualisations within a context that is relevant to the ecological discourse. Not to do so would exclude the contribution of Indian thought's unique conceptualisations of nature that can be understood from these divergent sources.

A possible question that one may ask here is about the divergent time periods of both Eastern and Western thought taken up for discussion in this book. Why does the summary of work in Western thought cover a range of conceptualisations from ancient Greek philosophy right up to the modern philosophers while the predominant discussions of nature in Indian thought cluster around pre-modern concepts? In this regard, I can only offer the explanation that the historical trajectory of a particular line of Greek thought lead to the development of natural philosophy in the West; in contrast, Eastern thought focused on refining language and debates. According to Comford (1991), the scientific tendency can be traced through a line of thinkers of the Milesian school and their successors such as Anaxagoras:

> These atomists succeeded in reducing *physis* to a perfectly clear conceptual model such as science desires, comprised of little impenetrable pieces of homogenous' matter,' with none but spatial properties—tiny geometrical solids, out of which all bodies, whatever shape and size, could be just built up (pp. 144–159).

Any philosophical theory and particularly moral philosophy is considered to be a highly abstract or theoretical analysis of ideas, and it may not directly result in any particular conclusions of practical significance. It is, however, equally possible those cultural presuppositions guide and inspire people in so many of the daily actions. According to Sasidhar (2006), the Vedic worldview with its "powerful symbolism" and "cosmic outlook" can provide an alternative vision which includes the animate and the inanimate aspects of creation. Human interests, beliefs, social structures, myths, laws, and many other aspects based on particular conceptualisations of nature have influenced the human relationship to nature in complex and variable ways. We can easily see this, for example, in the case of people choosing to be vegetarian or vegan from a sense of ethical responsibility to animals or humane treatment of animals. Though the choice is personal, there are some universal ethical aspirations behind the choice. Likewise, I suggest that influence of conceptualisations of nature within the Indian traditions of thought can be used to inspire people to relook at the relationship they have with the environment.

Environmental philosophers have expressed a lot of interest in alternative conceptual resources for understanding human response and relationship to nature within Asian traditions of thought. Some thinkers believe that the issues around the environment are due to Western ways of dualistic thinking. The unity of creation within Indian thought is often seen in opposition to this. The difference in the worldviews about nature influences our justification towards a positive or a negative attitude to conservation or use of nature. What kinds of concepts of the human being and concepts of nature are enumerated in philosophical traditions that would address the conservation practices in India? Further, one could also ask if traditions of Indian thought have successfully closed the gap and divide between the metaphysical categories of nature and the empirical practices of conservation.

Some thinkers believe that that the solutions to such problems must lie outside of Western conceptualisations. However, others such as Larson (1987) have warned that this kind of a "finding conceptual resources in Eastern thought" metaphor is similar to that of an orientalist/colonial project.[2] He claims, "If we were to substitute the word "natural" for "conceptual" in the expression "conceptual resources" it becomes immediately apparent that we are using an economic metaphor in our undertaking." Larson's question is valid in the sense that the so-called Western environmentalists look to Eastern resources for solving a problem created by exploitation of a non-human nature. I agree with Larson about the Orientalist project as far as the idea of Eastern resources to solve Western conceptual problems is concerned. But where does that leave me, a philosopher from the East? Am I a part of the orientalist project as a victim or perpetrator? How appropriate is it for me to try and apply the framework of the land ethics of Leopold to examine the complex ecological problems that India is facing currently?[3] I would like to clarify that the attempt in this research has been to try and keep to the role of an Asian philosopher, who is trying to look at alternate conceptualisations of nature that are valid in one's own social and environmental conditions within Asia in general and India in particular. The attempt is to try to articulate a culturally embedded ecology that rings true within the historical and cultural framework of India without delving into the debate about political origins of ideas and biases of culturalism. If conceptualisations of nature in the West have led to the idea of a separate non-human nature, it is not wrong to argue that a certain process of colonisation of Western knowledge systems and sciences might have altered the original paradigms, indigenous conceptualisations and presuppositions in the East, especially among the urban populations. Agarwal (2005), a prominent Indian environmentalist, speaks of his journey from "Western thinking" to understanding the deeper cultural history of his own country. He writes:

> The more I understood the way people lived with their environment, the better I understood their culture. … It was an intellectual journey that took me deeper and deeper into India and its cultural recesses making me despise more and more, as Jean Paul Sartre would put it, the "whitewashing" of modern Indians (p. 166).

5.2 Grounds for Oneness in Indian Philosophical Thought

As discussed in Chap. 1 of this book (also see Hargrove 1989, pp. xiii–xiv), one of the major issues in environmental ethics has been to rethink the theological basis of human–nature relationship. The ideas of oneness of nature in Indian Philosophy can be classified into three main themes. The first of these themes is based on the

---

[2]See also Chap. 11 for a discussion on Larson's critique of using alternative concepts from eastern philosophies to solve Western conceptual problems.

[3]For essays on the land ethics of Aldo Leopold, see Callicott (1989).

substantive oneness of nature across all creation or cosmos, a view that explains the unified origin and composition of the natural world. How is nature substantively one or seen as a whole? Another second stream of thinking explains the process of diversification and differences in the order of creation using different concepts such as the three guṇas of Sāṃkhya, different realms of beings, or combination of aggregates. A third view looks at the idea of a world of interrelations between all beings and the environment wherein a moral oneness is possible.

After analysing basic models of reality in Asian thought, Barnhart (1997) accounts for the Eastern idea of nature being linked to a concept of a graded reality. And he writes: "so, the qualified answer to the question of whether there is an Eastern idea of nature is yes, provided we understand that [1] nature means the way of being of a thing *in context* and [2] there are several contrasting accounts of the fundamental nature all things" (p. 424). He illustrates this view by explaining that in two pre-eminent schools of thought in India, Sāṃkhya and Advaita Vedanta; there is a clear distinction between the noumenal and the phenomenal. And he claims that of the two: "… it is the phenomenal that is of interest".

Elaborating this point, he mentions that both as māyā, the illusory appearance of the world in Vedanta, or as original substance *prakṛti* in the Sāṃkhya philosophy, nature as phenomenal existence, is not limited to the physical, but includes mental and intentional contents of the mind. Further, he reiterates that these traditions do not metaphysically set the humans apart from the rest of the existence. Bhattacharyya (2004) points out that the two streams of anthropocentric and eco-centric views of nature are combined into one stream of ecological thought In Indian tradition:

> In Indian Philosophy, these two interpretations of ecology are united. It has been repeated many times by the sages and seers of the *Ṛg Veda* that there is one universal consciousness which unites nature and man. Just as man without nature is an abstraction, so also nature without man is an abstraction. There cannot be universal consciousness restricted only to man or to nature (p. 4).

Here, it is important to also look at the critique of this unity as mentioned by Bhattacharyya from the perspective of ecology. In his essay of metaphysical implications of ecology, Callicott and Ames (1989) write that the unity of nature in Indian thought given by the philosophy of Vedanta is substantive and essential. The oneness of the Brahman or the consciousness is a distinctive type of oneness. This oneness according to them is "one in the manner that indivisible, homogeneous, quality-less substance is one". This is the difference which makes it difficult for a conceptual equivalent between ideas in holistic ecology and the Vedanta worldview. According to Callicott (1987), the holistic concept of ecology considers unity as "systemic and internally relational". Explaining this metaphysical idea further, he claims that within holistic ecology, nature is a *structured, differentiated* whole (emphasis mine). He writes: "The multiplicity of particles and living organisms, at either level of organisation, retain, ultimately, their peculiar, if ephemeral, character and identities" (p. 124). According to Larson (1987), while the idealist schools may not yield such a unity equivalent, it is the schools of Sāṃkhya and Yoga which have the concept of *prakṛti* constituted by the three *guṇas* as systemic and internally relational.

Boruah (2005) too emphasises that the unity equivalent has an implication to follow a moral order:

> "Whether in *Sāṃkhya* or *Advaita Vedanta*, the cosmocentric vision is intended to shape our place in reality and to regulate our conduct not with an exclusivist principle of human autonomy, but in tune with cosmic order" (p. 27).

Just like the idea of nature, *prakṛti* is also treated as a category "whole". *Prakṛti* is not a sum of its parts, *guṇa*s. The "whole to constituents" relationship is co-constitution rather than a sum. The Sāṃkhya dualism is very different from other kinds of dualism as seen in the paradigms of Western thought. Where "nature" (from a Western perspective) is incapable of serving purposes of liberation, the manifestation of *prakṛti* can be used by living beings for obtaining liberation. We find here again the link between the moral or transcendental realm of matter and the ideal of *prakṛti*. Sāṃkhya philosophy emphasises that *prakṛti* is the matrix of the whole psycho-physical universe. *Prakṛti* which is co-constituted by the *guṇa*s is defined by the process of evolution. As we have seen earlier, *prakṛti* as *pradāna* is the cause of all effects including the phenomenal world.

The idea of nature as being a creation is covered by a universal moral law or an order that determines a set of relations and consequences of activities is found in various doctrines of Indian philosophy beginning with the Pūrva Mīmāṃsa school. Dutta's (1936) paper is one of the earliest papers on the concept of nature. His views are significant as he breaks away from the Western concept of a homogenous "native" world view of nature that existed in writings on Indian thought and instead looks at the diverse equivalents of nature in the philosophical systems. His paper provides a brief background summary of the nature as a moral principle that is linked to the actions of human beings and other organisms. The moral principle *ṛta* as described in the Vedas and clarifies that from a law that governed planetary movements and the duties of the various gods, *ṛta* evolved into a principle of moral action and righteousness, its meaning taking on a similarity with "truth".

According to him, nature in Indian thought is conceptualised as a combination of "organic"—which consists of the embodied organisms and "extra-organic" components which is the environment—being governed by a single law that has been referred to in various philosophical schools as *ṛta*, *karma*, *apūrva* or *adṛṣṭa*. This law, according to him is a moral law that is impersonal, independent of a concept of a god and it maintains the order in the universe as an ordainer. The conceptualisations of nature in Indian thought are linked to the relationships between the human beings and the *loka* that they occupy both metaphysically and morally. The Indian philosophical traditions reflect this view. Dutta (1936) writes that the conceptualisation of nature in many Indian philosophical schools of thought is linked directly or indirectly to the idea of a moral law:

> With the exception of Cārvāka, or allied schools of materialism, any systematic account of which is not available, all Indian schools conceive of nature as the stage for moral beings,

> constituted and guided by moral needs. This holds good not only of the theistic schools but also of the atheistic ones like the *Buddha, Jaina and Mīmāṃsa* (p. 223).

Dutta (1936) also suggests that the conceptualisation of nature is not anthropocentric (he uses the term anthropomorphic) as the organisms bound by this law are not only human beings. The embodied organisms include a wide range of living beings starting from plants and ending with the divinities belonging to a category of organisms.

The philosophical school of Mīmāṃsā has a principle of *apūrva,* which Dutta (1936) considers as the moral principle that constitutes in the world. He explains that the Mīmāṃsā school used this principle to justify their central idea of ritual and action-yielding results accordingly:

> It (the Mīmāṃsā school) came to hold the theory of *apūrva,* according to which an action performed generates an imperceptible potency in the soul for the future enjoyment or suffering according to the merit of the action, and by an inexorable law this potency bears actual fruit in the future (p. 225).

Similarly, he describes that for the Vaiśeṣika school, *adṛṣṭa* was an indivisible principle, independent of God that leads to the formation of natural objects. The same principle in the theistic philosophy of Nyāya-Vaiśeṣika is interpreted as the principle of divine justice that rules over all the atoms in the universe. Dutta (1936) describes *prakṛti* as the material principle of the Sāṃkhya School and suggests that as *prakṛti* is teleologically connected to the "enjoyment of *puruṣa*", it is guided by a moral principle inherent within it.

5.3 *Guṇas* as a Framework for Moral Action

In Sāṃkhya philosophy, the conceptualisation of nature as *prakṛti* constituted by guṇa, gives rise to a special form of ethics that when extended to the framework of human treatment of environment allows for different gradation or norms that are context based.

As already explained in Chap. 3, the *guṇas* are regarded as the ultimate constituents of matter and primal objective entities. The *guṇas* are both the unifying and the diversifying principle that constitute nature as *prakṛti*. Sovani (2005, p. xiv) explains: "So the world is transitory distributively but not collectively". *Prakṛti* contains the opposing constituents—sattva and tamastamas—and yet they are reconciled by rajas. So the *guṇas* both create divisions and differences and still maintain unity. These as "substance cum evaluative" existents, provide a foundation for the internal relatedness amongst everything that "is". We can say that the *guṇas* in some way are essential, in the sense that they are the metaphysical basis of creation, but their expressions are more like attributes, which can be subjected to judgements.

The different objects of this world are endowed with different combinations of *guṇas*. Jacobsen (2002, p. 240) describes in his book: "Every phenomenon in the cosmos except the *puruṣa-s* is constituted by the three *guṇas* and every phenomenon can be experienced and is analyzable in terms of these constituents[4]".

It is important to understand that in every context or encounter, the dispositional *guṇas*, which are like attributes, can change. Again, we can say that *guṇas* form the relationship between the predisposition, knowledge and moral action (*karman*). Associated with action and its outcome, the *guṇas are* linked to rebirth and liberation, as well as the ability to change one's predisposition for the morally better. Jacobsen (2002) writes:

> The *guṇas* are associated with ethical values (i.e., *dharma* and *adharma*) which determine which kind of rebirth, with knowledge and ignorance which determine liberation and bondage, with detachment and attachment which determine transmigration, and with power and powerlessness, which determine degree of control (pp. 247–248).

The central point of the above argument is that it is only human beings who are capable of action as karma; all action is inspired by the dominance of one *guṇa* or the other, bringing all of the actions under the judgment of a moral kind.[5] This moral consideration has nothing to do with an external deserving of the object of moral consideration, but with the moral imperative, the predisposition—inner *guṇa*—of a moral agent. In fact, Sāṃkhya philosophy insists that except the *puruṣa* (consciousness), who has non-agency (*akartṛtva*), all other forms of neutrality or inaction involve maintaining the body in a state of *tamas* that falls under the agency of action (*kartṛtva*).

As in most Indian classical traditions of thought, though the idea of ethical behaviour has the ultimate purpose of liberation, still the norms and values related to correct behaviour based on one's place in the universe and context of one's life circumstances is encouraged. It is perhaps in the Mahābhārata, particularly in the *Bhagavadgīta*, that one finds a description of a detailed relationship between moral action and the *guṇas*.

5.4 The *Bhagavadgīta*: *Guṇa*-Based Action as an Ethical Framework

The *Bhagavadgīta* is not a classical Sāṃkhya text.[6] But one cannot ignore the fact that Chaps. 17 and 18 of this text are completely devoted to the description of the influence of *guṇas* on the ethical and moral action of human beings. From my experience of working on conflicts around the privatisation of a lake, I extend the concepts of the *guṇas* to an eco-ethical framework. Though largely a pragmatic

---

[4]For Jacobsen (2002), since the philosophy itself insists on liberation, the ethical analysis is better done at the level of *guṇas* than at the level of *prakṛti*.

[5]The lower realm beings such as the animals have no opportunity "not to obey" their *dharma*.

[6]Larson and Bhattacharya (1987, p. 8) refer to this text as a potpourri of various philosophies including cosmological and psychological Sāṁkhya reasoning and various other philosophies.

exercise, such a thought experiment perhaps is culturally similar to moral categories in Indian traditions. In this section, I use a reading of the *guṇas* from the *BG* to explain the relationship between an action and its effects, as well as norms for dealing with actions of others.

Most conservation discourse is about landscapes such as lakes, rivers, forests. Or it is about wide areas. In some instances, it is about a particular group of animals—whales or birds or tigers. Yet again, it is about natural resources such as coal or water. Sometimes, the issue about what is to be conserved becomes ambiguous. For instance, it can be asked as to what the lake is; is it a tank for water storage, a wilderness for water birds to nest, a backdrop for fun, or a source of water, fish and cattle feed? Some of these activities coexist, while others are in opposition. The washerman drying clothes in plain sight of visitors may not be right, but the birds may be able to share the space with the park visitors. While birds make it a nesting site, fishermen make a living from it, and the visitors find the lake an ideal place for recreation. All these encounters are in the form of some action, tangible, or intangible. The problems of conservation within the Indian framework are not about the human–non-human conflicts, but about the conflicts around the lack of relatedness with our environment and with each other as communities. When one calls a lake a "public space" and prioritises the public who use the lake as visitors more than the ones who use the fish in the lake, the way the lake is perceived is actually "private" or at the most, "elite public". Any kind of environmental action that focuses on any user does not have only a binary of ecologically sound or ecologically detrimental effect. The ecological cause that benefit some has mixed impacts, and this mixed impact can be captured best in the "fuzzy evaluation framework" of the three *guṇas*.

Here, the three *guṇas* are used as a moral framework for evaluating various kinds of activities, people, and objects using a Sāṃkhya-type of classification. The presupposition is that sattva and the development of its predominance is closer to the goal of liberation. Here, we find that the three *guṇas* have been used to describe the ethical rightness of an action by indicating that those actions performed with the predominance of *sattva* cause pleasure; those of rajas nature result in pain, and the actions performed under the influence of the tamas result in ignorance or violence. In the *BG*, Verses 20, 21, and 22 of Chap. 17 describe the act of charity evaluated according to the three *guṇas*:

> A sattvic gift is one that is given at the right place and time, with the thought that it is good to make a gift, to a deserving a recipient, who cannot make a return for it (Verse 20).
>
> It is said to be rajasic when offered unwillingly, with the expectation of a gift in return or with an eye to some advantage (Verse 21).
>
> A tamasic gift is that which is improperly and insultingly offered, at the wrong place and time, to undeserving recipients (Verse 22) (trans. Warrier 1983, pp. 541–543).

We should infer that according to Sāṃkhya, no action is free of any *guṇas* at any time. Even the most noblest of sattvic action will also be mixed with the other two *guṇas*. The lake development contractors wanted to fence the lake to prevent encroachment by petty shops and private landowners around it. Consider that this is a *Sāttvika* action. The fencing, however, also allows them to collect entrance fee from all the visitors and make some profit from it. This causes pain to some poorer class of visitors who may not be able to afford the fee. These two are *Rājasika* actions and impacts. To others such as the weed collector or the dhobi, their entire means of livelihood from this lake is wiped out. This is the impact of *tāmasika* part of the action.

The *guṇa* accounts in the *BG* suggest that the predominance of certain *guṇa*s at certain times influence behaviour and action. On the other hand, certain actions themselves induce the development of particular *guṇa*s. Through this explanation, the gap between attitude and action, disposition and behaviour (in Indian thought) becomes relationally cyclical. *Guṇa*s become a bridge between the internal mind-intentions and actions performed in the external world; they are the link between the action and its effects and between beings and their worlds.

This ethical framework is interesting because though in some sense the *guṇa*s are constituents of the human being, a person has the ability to change the predominant nature using free will and actions. In other words, a *rājasika* person can also perform an action that is *Sāttvika*. By repeated performances of such acts, the internal nature becomes *Sāttvika*. In contemporary terms, when an act or a phenomenon of experience with nature occurs, there is the production of a specific narrative, a kind of making sense, or a symbolisation creating sites. These narratives, whether the narrative of development or the narrative of the sacred waters goddess or the narrative of conservation, are all value-laden in some sense or the other. The conceptualisations of the lake are not value-free. Our experiences are a confluence between conceptualisations and concrete phenomenon, where we try to make sense of the world through a notion of place and movement. Narratives are constructed from values and give rise to other values in return.

Human interaction with the environment cannot be evaluated in terms of a binary of no interaction or over-exploitation. The *triguṇa* perspective gives us a related framework to evaluate our actions in a graded manner. We can say that it gives us a framework for weightage of values rather than a mere moral significance of them. Extending this *triguṇa*-inspired action hypothesis to the environment, one can say that those actions which are about the protection of the environment (we can call them "eco-friendly") and lead to pleasure, joy, and good of all concerned are *Sāttvika*. These would be actions performed with responsibility, restraint, and care, which would lead to a fulfilled life on our planet. Those actions that are exploitative towards nature, purely aimed at creating maximum economic benefits at the cost of other beings are *rājasika*, and such actions would lead to short-term pleasure and long-term pain. Human actions that are today popularly called "development" would be *rājasika* in nature. Building dams on rivers, mining, and utilisation of natural resources are all human enterprises of economic profit and gain. Tamasic actions would be those that cause wasteful use of nature or employ violence, and are inspired by misconceptions and ignorance. These result in both needless destruction and annihilation. The categories of pollution and the destructive machinery of war come under *tāmasika* action, which causes only distress to all concerned.[7] From the injunctions given in Indian philosophy, *sattva* is to be nourished, *rajas* is to be bound, and *tamas* is to be destroyed. So the responses to the actions around the environment can be articulated as follows: *sāttvika* activities around nature can be encouraged, and *rājasika* activities have to be regulated and controlled by policy and law. *Tāmasika* activities should be totally banned or replaced by alternatives and punitive action should be taken against the perpetrators of such acts.

So the idea of fencing a lake to prevent encroachment though causing pain to some (through *rajas*) is still beneficial to the lake, the birds, and the park users, who feel safer inside a guarded area. Illegal dumping of construction debris and garbage and sewage has to be stopped (a *tamas,* predominant act). While the dhobi lost the use of the water completely, the cattle feed collector was allowed to cut grass early morning before the fee-paying visitors arrived (activity slightly *sāttvika*—to feed a cow—but mainly *rājasika*). Daily morning walkers had no entry fee (*sāttvika* activity for health), while those who came for "entertainment" at other times (predominant *rajas*) paid a fee.

The problems with current evaluation without this framework is that the there is no gradation between "using nature", "exploitation of nature", and "destruction of nature". What this new perspective does is to give us a practical method to evaluate our actions in a graded manner and therefore avoid the binary of either right or wrong, which is a crucial requirement in environmental ethics. The activities themselves are not right or wrong morally but their effects are. Since the action is judged by its final effects as well as its predominant intention, it also does way with the problems of a mere right and wrong binary.

5.5 Human and the Non-human: Sameness and Difference

As discussed in Chap. 1 of this book (also see Hargrove 1989, pp. xiii–xiv), one of the major issues in environmental ethics has been to rethink the theological basis of human–nature relationship, particularly problematic is the division between the human being and the rest of the beings (human–non-human divide). Ethical discourses of the environment often centre on discussions of inclusive moral frameworks for animals and other non-human sentient beings. As discussed earlier, the difference between humans and non-humans not only delineates "nature" from "human", but it also has implications for the moral philosophy. Some questions that are discussed and debated are as follows: how do we position human responsibility towards other animals? Do animals have rights? These questions centre around two broad themes. First theme is about the nature of animality in human beings.

[7] I can use the analogy of the fishing practices to further clarify my point. To fish and throw back the fish into the river would be *sāttvika* (for conservation). To fish and take it away for dinner would be *rājasika* (soon the resources will be depleted). To electrocute the water and kill all the fish and pick up the dead floating fish at one instance would be *tāmasika*.

What are those characteristics that are common to all animals? Secondly, the debates focus on the differences between the human–animal distinctions (Calcaro 2008). Following some traditions and other themes in Indian philosophical thought, a few of the conceptual frameworks that shed some clarity on the divisions between the human–non-human in Indian thought will be taken up for discussion here.

Soper (1995, pp. 37–38) points out that the difference between humans and the rest of nature is dependent on humanity's ability to create both natural and artificial products through rational deliberation. Another of these abilities is to follow convention and exhibit behaviour that is in line with such norms. She suggests that historically such a dichotomy between human beings and the rest of nature has been common in Western thought:

> An opposition, then, between the natural and the human has been axiomatic to western thought, and remains a pre-supposition of all its philosophical, scientific and moral and aesthetic discourse, even if the history of these discourses is in a large part a history of the differing constructions we are asked to place upon it (p. 39).

The categories and terms used to describe the universe and its components and experience of the world in Indian thought are distinct from Western philosophical conceptualisations of nature as "non-human". The dichotomy of human beings versus the rest of nature is not a rigid value designator. Indian thought, however, is not completely devoid of such a distinction either. One of the poems of "good words" (*subhāṣita*) composed by the Sanskrit poet Bhartṛhari states rather audaciously that a person with no cultivation of music, literature, or the arts is verily like a domesticated animal, a burden on the earth, only different from the animal in not having a tail and horns and not eating grass.[8]

Soper (1995, p. 41) draws our attention to the philosophical issue that is at the root of this problem. She says that it is not the idea of a difference between the human and the non-human that is important but "… the way in which it is to be drawn, and more importantly whether it is conceptualised as one of a kind or degree?"

One of the ways in which the above distinction is posited is the idea of nature as an object of experience for the human subject. Is there such a train of thought in Indian philosophy? Soper describes this subject-object division thus:

> In modern philosophy, the humanity–nature antithesis has on the whole been broadly and abstractly as a Subject–Object division: as a division between the Humanity that is the thinking subject and the nature that presents itself to thought, but is incapable of thought itself (1995, pp. 41–42).

Such a Cartesian divide is likely to strongly support the distinction between the human and non-human based on rationality and other arts as in the case of the poem we have seen earlier. But this is not so typical of Indian traditions of thought. As we have seen in chapter three, the world of creation is often classified as a knowable

[8]Translated by the author from Sanskrit quoted in *Bhartṛhari*, *Niti* and *Vairagya Shatakas* (Joglekar 1911, trans., pp. 2–3).

world in Nyāya thought. There is also a distinctiveness articulated in the 'order of beings' framework in Sāṃkhya. Both of these philosophies in their own way seem to imply a Cartesian divide. However, a metaphysical analysis of the knower and the objects of knowledge in Indian philosophy would clarify the nature of this distinctiveness. Let me highlight some of these ideas from the *Bhagavadgīta* (henceforth, as *BG*), one of the key texts common to the Vedānta schools on one hand and deeply connected with Sāṃkhya metaphysics on the other.

In the 7th and 13th chapter of *BG*, we find a discussion on what seems like a clear division between *prakṛti*, the field called *kṣetra* and the knower of the field, the *kṣetrajña*. One interpretation of this is that the field, the object is nature, and the knower is the subject. But a closer reading of the other verses in the *BG* gives us a different perspective. The knowing subject in this case is not the experiencer of knowledge per se, but the consciousness principle provides awareness for the very process of thinking. The *BG* (Verse 7.5) describes the creation as a result of the union between two kinds of nature that belong to The Lord, the higher immutable and the lower mutable. The next verse further states that: "All things in the world born moving or unmoving are from the union of the field and its knower". Such a creation that is formed by the union of *prakṛti* (nature principle) and *puruṣa* (the witness principle) is pervaded by the divine in all its aspects. Śankara in his commentary on the text points out the field stands for all of creation beginning with the human body. The *BG* goes on to list mind, intellect ego, the great five elements, organs of sensing and the mind, organs of action, and objects of the senses within the category of the field. On the other hand, the knower is the singular supreme knower who has control over the field, in the sense that the mutations of the field do not have an effect on it.

The dualism articulated here and in many other similar texts is that of the divine god (supernatural) versus nature (creation) where god becomes the support of the creation. Given such a conceptualisation, it becomes difficult to imagine a human–nature duality within the concepts of the field and its knower. Nature's existence (which is inclusive of the thinking apparatus and ego) and the experiences of a human being occur within the background of existence provided by the supreme consciousness. This dichotomy is more of a witness–experiencer type in case of an individual self or a supernatural–natural type in case of the created world. The mutable–immutable origin of the universe actually equalises the created beings instead of separating them.

If this is the conceptualisation of human–nature in these traditions, particularly where thinking does not play a major role, how then are we to understand the differences between the humans and non-humans such as the animals? It is obvious that even in a naive realistic way, there ought to be some kind of difference between humans and animals at least? Humans seemed be placed apart from other created beings and objects. What then is the basis of such a distinction if it has nothing to do with the mind?

There are two main hypotheses I posit here that may clarify the kinds of conceptual differences we ought to look for. These two themes are only illustrations of the kinds of separation possible between human and the non-human from two different streams of thought. The first one is based on theme of the four human purposes called *puruṣartha*. This discussion will examine in brief the narratives of teleological differences between human beings and the rest of nature. The second explanation is related to the body–body difference that is so ingrained in the theory of karma. Both these discourses of *puruṣartha* and *karma* are tied up with moral claims and ethical action and provider alternative ways of looking at the relationship of the human and the non-human.

5.6 The Goals and the Body: Divergence of Human and Non-human

*Puruṣartha*s refer to the purposes that give meaning to human life. This theme of Indian traditions perhaps is an important concept that helps us find the point of divergence between human beings and the rest of nature. Human life and activities are envisioned through a framework of broad teleological categories that classify all human endeavours into four kinds. These are *dharma* (righteousness, duty), *artha* (wealth and livelihood), *kāma* (desire) and *mokṣa* (liberation). These four categories are not actions in themselves but represent the broad sphere under which all of human actions can be included. According to Mohanty (2000), these goals "… are to be pursued de facto by persons who are embodied souls living in the world and members of a community (i.e., who beings-with-others and beings-in-the world)" (p. 70).

Dharma, one of these four goals, is the righteousness that is given by one's place in the universe. The *dharma* that is determined by one's form, role, function, and innate nature is common to all created beings and even the world of materials. In this sense, *dharma* is almost like a natural law. Each created being is almost impelled to act as per its *dharma* to maintain order and stability in the creation. The *dharma* of rivers for instance is to flow downwards and that of fire to burn. The *dharma* ensures members of groups and community follow and act appropriate to their group's position and status in society. While the class of other beings have the most basic of the *dharma*s given by their instinct and innate nature, human *dharma*s are much more complicated bound to moral choice and action. Each community and its members are bound in a web of relations and functions that are multi-layered. For each person, there are *dharma*s that are to be followed of being human, being female, being a daughter, being from a particular caste, being from a place. Needless to say that the choice of what *dharma* to follow when is contextual and often leads people to face moral dilemmas—a choice—between two equally appropriate courses of action.[9] The other conflict or ethical choice, that is to resist one's dharma or to choose a course of action rests only in the human condition. The rest of the non-human world is impelled to follow their *dharma* by either a sense of

[9]For a detailed discussion on moral conflict and moral dilemma within *dharma*, see Matilal (2002, pp. 92–93).

deep responsibility such as the supernatural beings or by being bound to their inner nature, such as the plant and animal beings or, being created to fulfil functions by divine or creation ordinance, such as *avatārs*.

The other three goals—*artha* (wealth and livelihood), *kāma* (desire), and mokṣa (liberation)—are *puruṣarthas* that are unique to human beings. These three goals categorise human actions into broad spheres of human activities. To be morally appropriate, these three spheres are bound by the limiting injunctions of *dharma*. In a way, dharma forms the moral limits of free will represented by desire, livelihood, and spiritual aspirations. It is believed that dharma sustains the natural order of the universe. For instance, *kāma*, the desire for sensual experiences, is to be directed towards appropriate objects. While desire for one's own wife is appropriate, coveting another person's wife is discouraged. Similarly, the means to acquire wealth is determined by one's position and role in society. In fact, literature even indicates that even thieves had to follow a particular *dharma*.

The other beings do not have access to the *telos* of the *puruṣarthas*. It is these ends that make human life distinct from the lives of all other sentient beings. The theme of *puruṣarthas* describes the human condition, but does not completely explain the causal processes of unequal divergences. How does the *kṣetrajña* as the inner self of all beings acquire this variation of manifestations? In other words, how are humans born humans while other souls are born animals? This is explained using the doctrine of karma and its effects. The chance one can be born in an animal's body or a human body is a possibility created by merits and demerits (*punya* and *pāpa*), and therefore, the texts such as the Upaniṣads also call these "lower wombs". Within philosophical traditions that believe in transmigration of the soul or rebirth, the human body is a moral attainment that is the result of good deeds performed in previous lifetimes. Those with karmic deficit are born as animals or non-human creatures. According to these beliefs, the animal bodies, such as other "inferior" bodies such as those of women or those born in the unprivileged castes are the sites for experiencing the effects of some moral transgressions performed in the past. Both caste- and gender-based *dharmas* have been a source of contention for today's understanding of *dharma* as an ethical category. This is an important area of critical enquiry. This answers the question is about the moral adequateness of *dharma*, and the way it is construed to differentiate between the privileged human and the non-privileged humans (also called subhuman in some literature) or non-human. The privileges of being born in particular sections of society are explained according to a system of naturalised hierarchy, supported by the framework of *karma* as causal and *dharma* as limiting rights and privileges.

Karma becomes causal in explaining the very deterministic hierarchies in Indian socio-cultural traditions (Baindur 2014). To be free of the effects of such moral retribution in the body is impossible within a physical body that is already bound by birth into a position in an order of hierarchy, pre-determined socio-religiously for both natural and social classes. Therefore, somewhat equally created souls are born in animal bodies because as humans in previous lifetimes they did not fulfil their moral obligations. The restitution of such transgressions cannot be made if one has an animal body because it is only the human body that can act morally.

Moral agency is limited to the human sphere of activities. Within this discourse, only human bodies with the ability to perform intentional action can modify their positive or negative balance of *karma*. Human beings also are bound to moral agency through the sphere of *dharma*.[10] Non-performance of assigned *dharma* can also lead to negative effects. On the other hand, *karma* for the non-humans such as animals is merely activity based on their innate nature, particularly within the Veda-based philosophies.[11]

This karmic discourse of body–body difference is perhaps the most difficult to adapt to an environment ethic. But it still could form the foundation for expressing the virtue of compassion. The ethics is still normative, yet this is perhaps the only way to create some moral standpoint from the divergent body narrative. The fact that the non-privileged bodies are already in a state of retribution and in a way experiencing the effects of their previous karma, it is important for us who are human to follow the *manuṣya dharma*, the human duty of being compassionate to all beings.[12] Time and again in the Mahābharata, we are asked not to needlessly hurt or kill other beings—*ahiṃsā paramo dharmaḥ*—abstaining from violence is the best duty (Das 2009, p. 49). In a later chapter, we will examine the concept of *ahiṃsā* from a broader perspective of different philosophical streams and understand its ethical implications.

5.7 Human and Non-human in the Order of Beings

Given the variety of beings in creation, a class categorisation of these as collectives is perhaps the easiest way to differentiate between human beings and nature. We have seen that a metaphysical classification of nature relies on irreducible wholes like *dravya* or *padārtha*. On the other hand, a naive conceptualisation is based on a certain world view of "beings and the metaphysical worlds" they live in—that one can call a *loka*-centric view of the universe—is prevalent in the belief systems of many traditions in India. According to Indian cosmology, the cosmos consists of many worlds called *loka*s (which are sometimes mythical and not *on* the earth) and also beings (*bhūta*s) that inhabit such worlds.[13] Kinsley (1995) elaborates on this perception, when he explains: "In many Hindu scriptures, it is clear that the world is perceived as being alive with forces, powers, spirits, and deities that express themselves through what we call natural phenomena". All elements of a cosmos find a place in a complex hierarchical humanised cosmic order in Indian thought.

---

[10]Beings such as gods and other-worldly beings have their own injunctions of *dharma* and *karma*.

[11]The Jainas have a different theory that is addressed in the forthcoming section.

[12]The critique of this position is that it may result in human beings developing a moral high stand about being superior. As this hypothesis also applies to caste and gender issues, the dangers of this compassion turning into a newer form of oppression or high-handedness are possible.

[13]For details on some of these sacred beliefs, see Chaps. 7 and 8 of this book on themes of nature as sacred.

We find a philosophical explanation of beings in the enumerative philosophy of Sāṃkhya.

Larson and Bhattacharya (1987) quote a cosmological myth of evolution and the properties of various realms of beings as described in the Sāṃkhya text, *Yuktidīpika*. According to this myth:

> The great being (one of the great bodies or *māhātmya śariras*) at the beginning of the world cycle had no offspring to carry on his work (Karma). He created five 'mainstreams' (*mukhya strotas*) that are characteristic of the plant realms through meditation but they were insufficient. Similarly, he then created a set of 28 horizontal streams called *tiryakstrotas*—the characteristics of the animals, birds and insects—but that did not accomplish his desire either. The upward moving streams—characteristic of the divine—that were created afterwards did not also serve his purpose. Finally, he created the downwards streams (*arvaksrotas*) characteristic of the human realm that fulfilled his needs. Therefore it is the human realm that accomplishes action for the great being through Karman (p. 58).[14]

According to *Sāṃkhyavritti*, sarga, or creation, is threefold: the manifest world is created by *tattva sarga* which is essential or elemental creation, *bhāvasarga* or the creation of predispositions and finally *bhūta sarga* or the creation of the gross universe. What we see around us as the natural world is therefore only one kind of creation, the gross world or nisarga also known as *bhūta sarga,* the downward evolute of primordial nature, *prakṛti*. The empirical world of *bhautika sarga* is described as consisting of different beings of higher, middle, and lower orders. The higher, divine, order has five varieties of beings, the lower order is of fivefold types. The humans in the middle order are only of one kind (Larson and Bhattacharya 1987).

In the created world, called *bhautika sarga*, the derivatives of *prakṛti* are also classified according to the *guṇa*s. The phenomenal creation is summarised by Larson and Bhattacharya (1987) as follows:

> The projective force of the fundamental predispositions, together with the subtle body, generates not only the human realm but also an eightfold divine or cosmic realm and a fivefold animal and plant realm. Taken together, the projected realms are referred to as the external world (*bhautikaasarga*) with *sattva* predominating in the divine realm, *rajas* in the human realm and *tamas* in the animal and plant realm (p. 26).

Let us discuss the idea of moral considerability, which has been addressed as an issue in environmental ethics. Taking a cue from Goodpaster's (1978) essay, if we ask the question: "what is the moral considerability of nature?" interesting issues can be raised about the framework of Western environmental thought. He asserts that the term "moral considerability" is different from the issue of "rights":

> My inclination is to construe the notion of rights as more specific than that of considerability, largely to avoid what seem to be unnecessary complications over the requirements for something's being appropriate "bearer of rights" (p. 311).

---

[14]Summarised quotation from Larson and Bhattacharya (1987, p. 58).

In summary, we can say that philosophies that hold a life-centric theory (or sometimes called bio-centric theory) of environmental ethics support the idea that the moral treatment of nature is based on the inherent worth of nature, otherwise called intrinsic value. Human-centred theories are those that are founded on the idea that our moral treatment of nature is ultimately based on the well-being of humans. Both these theories are possible within a framework that admits of the differences between the human and the non-human world.[15] Even the bio-centric view of nature can take into account some sort of moral considerability of nature by the human beings. Indian thought, as we have seen so far, however, does not admit of this kind of a clear duality. Within the Indian thought, nature as *prakṛti* is cosmic and is inclusive of all created or existent components. Therefore, it would be a contradiction to speak of a separate "environmental or nature ethics" within the field of Indian moral philosophy. So we need to find a different framework for an ecological ethics within Indian Philosophy. From the perspective of the Sāṃkhya narratives of creation, it is possible to distinguish between the human realm and the other realms, such as those of plants and animals using the classification under nisarga or bhautika sarga. Then, the following arguments for ethics are possible within the Sāṃkhya world view, related to the human world:

1. From the classificatory view, the human world is dominated by *rajas* and/or is capable of moral action (karma) and has attributes of *sattva* such as sentience in the form of discriminatory awareness (Jacobsen 2002).
2. The plant and animal worlds are dominated by tamas mainly and attributes of *sattva as* sentient existence (Jacobsen 2002).
3. Moral sphere of action (the agency of action) is of a *sattva–rajas* predominant kind.
4. Plants and animals being *tamas* predominant cannot act morally. They are subject to moral considerability, but not to moral agency.
5. Therefore, it is only in the human world that one can perform moral actions. These actions are further guided by different tendencies and predispositions of the *guṇa* that is predominant in a person at the time of action.
6. Moral development consists of not just right actions but development of those *bhāvas*—innate strivings or predispositions of living beings—that induce right action (*dharma*).

The idea of dharma therefore becomes a fundamental principle of ethical action that creates a relationship among all beings in Indian thought. The discussion on *dharma* continues in Chap. 8 on nature in Vedic thought. In this next section, we examine the Jaina concept of living beings that is the basis for the life-affirming ethics of non-violence, *ahiṃsā*.

[15]For a summary of these two views, see Taylor (1986, pp. 10–14).

5.8 Jaina Beings: Sameness and Difference

The Jaina world view is also a *loka*-centric view of the universe, given by geographical categories. The important difference of the Jaina school from the other traditions of philosophies is that it does not admit a creator or a prime cause for the universe. It is atheistic theologically and also in not admitting the authority of the Vedas. For this school, creation is *saṃsāra* consisting of cyclical periods of ages in which beings undergo enjoyments and suffering due to the accumulation of karma matter on their souls. Also central to the positions of beings in the Jaina universe is the doctrine of *karma* that determines one's position in the universe and the balance of pleasure and sorrow is determined by one's location in the soteriological landscape linked to their geographical position on Earth.[16] Jaina cosmography also delineates from some parts of the cosmos where *karma* has effect and also can be cleansed from the soul called karmabhūmi. Jaina geographical texts are meticulous in describing the universe with measurements in this regard.

At the outset, this seems somewhat similar to the views of Vedic philosophies, yet the Jaina idea of *karma* as form of matter that clings to the soul is different from the abstract notion of *Karma* that other philosophies of Indian traditions seem to share. True knowledge about liberation is realised, and one gets the understanding and proper perspective when taught by *tīrthankaras* (literally, the "bridge-makers"). These are souls who are not merely enlightened but also those who help other souls to reach the completion of their own journey to . The describes worlds ordered in morally significant categories, the beings migrating from one body to another located in the middle worlds. The beings that are the carriers and experiencers of the karmic world are called jīvas, in contrast to the non-beings, ajīva. But as souls, all beings share life, a common ground.

Even elements such as water, earth, air and fire, conceptualised as bodies of animals are also imbued with life or awareness, or *caitanya*. This is different from saying that the elements themselves are alive. These elements such as the physical air are *ajīva* or non-living components of the universe. The non-living parts of the universe are five extended substances (*pañcāstikāyas*): substances (*dravya*), *ākāśa* (space), means of movement (*dharma*), means of rest (*adharma*), time (*kāla*) and matter (*pudgala*). The living components called *jīvas* are souls which have energy (*vīrya*), *caitanya* (awareness), and *sukha* (bliss). *Jīvas*, as unenlightened beings, are involved in the infinite wheel of pain and suffering, till their liberation.

Therefore, the Jaina classification of beings begins with the distinction between liberated and non-liberated beings. The liberated beings exist in disembodied state, free from suffering. The non-liberated embodied beings are of interest to us to understand the Jaina views of human and non-human in nature. It is very clear that the common fate of all souls is to transmigrate and be embodied in the body of any of the beings that exist upon earth. Pániker (2010) describes the Jaina world view as a relational universe:

[16]This will be taken up for discussion in Chap. 6.

> The cosmos is conceived in relation to the organisms that dwell therein; Thus rather than a coldly-static cosmos, we have to do with a living universe, with a living super organism. It is not the physical description which counts, it is the destiny of human beings and other beings (p. 38).

There are four fates for all souls to attain different kinds of bodies by the doctrine of *karma*: as humans, hell beings, divinities, or as animals and plants (*Tiryañca*) (Wiley 2006, p. 39).[17] Jaina philosophers insist that humanness is not valued above other forms of life. Human beings are unique in having the potentiality of understanding liberation or of having the extra-sense rationality above the five senses.

In fact, there are possibilities of animals, plants, and other beings that can be influenced by sacred doctrines and achieve spiritual progress. Umasvasti, the Jaina philosopher, regards all beings that are not human, divinities, or denizens of *naraka* (hell) to be animals. So, all animals and plants are not only non-human, but also non-divinity and non-hell beings (Pániker 2010, p. 48). The vertical erectness of the human body during movement and divinities seems to be the criteria of difference in naming the animals *tiryañca,* those that move horizontally. The primary classification of beings into moving (*trasa*s) and non-moving (*sthāvaras*) is significant and occurs in many older Jaina texts (Pániker 2010, p. 48) and can also be regarded as a sub-classification of *tiryañca.*

Alternately, Jaina beings are also classified on the basis of sense organs they possess, and the way they experience the world—one-sense, two-senses and so on—humans having five senses in all. The most interesting of the forms of life that the Jainas describe in their classification is microscopic organisms or life-forms that have elemental bodies, creating a world or an environment that is thick with living beings. These smallest units of life for the Jaina are minute colonies of sentient beings, called *nigoda*s. These are called *ekendriya*s or one-sense organisms based on the sense classification of organisms. In the hierarchy of spiritual development, their souls are opaque due to the dense *karma* matter and they have not even started on the path to liberation. It is important to note here that the pathway to liberation is not a gradual or linear movement from one-sense to two-sense bodies and higher forms of bodies. Every life form has an equal chance at the four destinies. The *nigoda*s are eternal and may be reborn in other bodies if a human soul attains enlightenment, or if a human is born as a *nigoda* due to *karma*. Unlike Vedic systems of thought, the soul need not linearly work its way up through experiencing the *karma* in different bodies (Wiley 2006, p. 40).

The *ekendriya*s that occupy water, earth, or air, bodies differ from each other by the variation in a kind of *karma* called *āyukarma*, which gives the longevity of any being and determines the type of embodiment of the next life. The water-bodied organism lives for 7,000 years, while fire-bodied beings may last only for three days (Wiley 2006, p. 40). While the one-sense organisms develop vital energies called *prāna* and also the strength of the body and respiration, they only have the sense of

[17]The *svāsthika* as a symbol represents the four fates and is revered by the Jainas (Pániker 2010, p. 48).

touch and ability to produce sounds. The two-sense beings develop the sense of taste. The two-sense beings can exist as single entities or as colonies. Many of the vegetative beings come under this category. Trees, grasses, and bushes come under the single *jīva* category, while figs, potatoes, onions, garlic, and so on are *jīva*-colonies (Pániker 2010, p. 54).

The three-sense beings have all the previous characteristics with the additional sense of the smell. The sense of sight is added to the four-sense beings and that of hearing to the five-sense beings. The humans in addition to the five senses have the rational sense and are referred to as the rational sense beings, *pancendriya saṃjñīs* (Wiley 2006, p. 41). However, the idea that rationality is limited to human beings is often contradicted by accounts of animals that act morally and listen to scared teachings and move onward to liberation. Pániker (2010) recounts that even Mahāvira in one of his earlier births was born a lion was instructed in non-violence and after many other lives ultimately reached the state of a *tīrthankara*.

> Even though they are one-sensed beings, these minute life forms with earth, water, fire, or air bodies also interact with the environment and take nourishment through the surface of their body. They have instincts that cause them to fear, to reproduce, to gather things for themselves, and to also reproduce (Wiley 2006, p. 41).

Jaina texts explain that with only the sense of touch, these beings can cause actions that are subject to the various passions. Quoting from the scriptural references, Wiley (2006) explains how the beings have volition: "It is said by merely breathing, earth-bodied, water-bodied, and fire-bodied beings, as well as plants, commit three, four or five types of actions, while an air-bodied being, stirring part of agree or causing it to fall down, also commits three, four, or all five actions" (p. 42).

It has also been taught by Mahāvīra that these one-sense organisms also experience suffering through their sense of touch but in an indeterminate way. This idea lays the ground for the supposition that all beings experience pain or suffering even if they are microscopic or unaware.

Jain taxonomy of beings lists many beings and 8,400,000 possible types of birth are mentioned (Pániker 2010, p. 56). This list includes both commonly observed beings and also mythical beings and supernatural types. The extensive classification of the types of beings in the Jaina philosophy may seem like a descriptive taxonomic exercise, until we understand that this classification is deeply related not only to doctrine of *karma*, but also related to the practice of *ahiṃsā*. He writes,

> … texts are explicit in recommending the intellectual study of living beings in order to understand what ahimsa means and be able to practice non-violence leading to *Nirvāna* (*Sūtrakṛtānga* 1.11.11) When one knows where beings come from and where they go, when one grasps the mechanism and structure of *samsara*, one can take steps to escape it (from Padmanabh 2000, quoted in Pániker 2010, p. 56).

Pániker (2010, p. 273) insists that in this philosophical tradition, despite the varieties of beings enumerated, all souls are identical in nature, "there is no difference between the soul of a *Jīna* and that of a grub". What makes these beings different is the association of karmic matter that is bonded to the soul, through the medium of passions. Long (2011, p. 163) explains that passions are likened to the wetness of a cloth that attracts dirt. Practices of Jaina philosophy are directed towards burning of this *karma* that is possible in a human birth, which is cosmographically located in the area where both the (traditionally understood) chronological period and the geography are conducive to liberation.

From another perspective, the idea of nature that is in a sense replete with so many beings and particularly the need to conceptualise the micro-beings may have some metaphysical basis. One could suggest that the idea of movement represents the idea of volition for the Jaina philosophers. We can see a connection between life and the need to have vitality in elements to move or propel themselves (somewhat similar to an Aristotelian view). It is likely that in order to find a cause to explain movements of the air, water, earth, or wind, the theory of vitality in everything that moves results in a hylozoic framework. This is furthered by the absence of a divinity or supreme god who causes everything to move and function; hence, the Jainas prefer to look at the universe as organically endowed with life and passions that cause *karma*. This becomes an important basis for their practice of *ahiṃsa* (non-violence). As an ethical category that is found as a norm in many other philosophies, it will be taken up for discussion in detail in Chap. 11.

References

Agarwal, A. (2005). Can hindu beliefs help India meet its ecological crisis? In C. K. Chapple & M. E. Tucker (Eds.), *Hinduism and ecology: The intersection of earth, sky, and water* (pp. 165–182). New Delhi: Oxford University Press.

Baindur, M. (2014a). Nature, body and woman: An Indian perspective on value dualisms. In S. Sarukkai, G. McOuat, & J. Kanjikarrat (Eds.), *Sciences and narratives of nature: East and west*. New York: Routledge (forthcoming). Also 'Nature, body and woman: An Indian perspective on value dualisms,' during the *International workshop on sciences and narratives of nature: East and west*, December 12–14, 2011, Manipal University, Manipal (co-organised by Manipal Centre for Philosophy and Humanities and Situating Science Strategic Knowledge Cluster of Canada).

Barnhart, M. G. (1997). Ideas of nature in an Asian context. *Philosophy East and West, 47*(3), 417–432.

Bhattacharyya, S. (2004). Where Western idea fails. In D. P. Chattopadhyaya (Ed.), *Self, society and science: Theoretical and historical perspectives*, Vol. 11.2 of History of science, philosophy and culture in Indian civilization (pp. 3–8). New Delhi: Centre for Studies in Civilisation.

Boruah, B. H. (2005). Environmental wisdom. In D. C. Srivastava (Ed.), *Readings in environmental ethics, multidisciplinary perspectives*. Jaipur: Rawat Publications.

Calarco, M. (2008). *Zoographies: The question of the animal from Heidegger to Derrida*. New York: Columbia University Press.

Callicott, J. Baird. (1987). Conceptual resources for environmental ethics in Asian traditions of thought: A propaediatic. *Philosophy East and West, 37*(2), 115–130.

Callicott, J. B. (1989). Conceptual foundation of the land ethic. In *Defense of the land ethic* (pp. 75–100). Albany: State University of New York Press.

Callicott, J. B., & Ames, R. T. (Eds.) (1989). Introduction. In *Nature in Asian traditions of thought*. Albany: State University of New York Press.

Comford, F. M. (1991). The scientific tradition. In *From religion to philosophy: A study in the origins of Western speculation* (pp. 144–159). Princeton: Princeton University Press.
Das, G. (2009). *The difficulty of being good: On the subtle art of Dharma*. New Delhi: Allen Lane (Penguin group).
Dutta, D. M. (1936). The moral conception of nature in Indian philosophy. *International Journal of Ethics, 46*(2), 223–228.
Goodpaster, K. E. (1978). On being morally considerable. *The Journal of Philosophy, 75*(6), 308–325.
Hargrove, E. (1989). Introduction. In J. B. Callicott & R. T. Ames (Eds.), *Nature in Asian traditions of thought* (pp. xiii–xiv). Albany: State University of New York.
Jacobsen, K. A. (2002). *Prakṛti in Sāṁkhya-Yoga: Material principle, religious experience, ethical implications*. (1999 Reprint) New Delhi: Motilal Banarsidass Publishers Pvt. Limited, Indian edition.
Joglekar, K. M. (trans.) (1911). *Bhartrihari, Niti* and *Vairagya Shatakas* (with notes translation and critical introduction). Mumbai: Oriental Publishing Company.
Kinsley, D. (1995). *Ecology and religion: Ecological spirituality in a cross cultural perspective*. New Jersey: Prentice hall.
Larson, G. J. (1987). Conceptual resources' in South Asia for 'environmental ethics' or the fly is still alive and well in the bottle. *Philosophy East and West, 37*(2), 150–159.
Larson, G. J., & Bhattacharya, R. S. (Eds.) (1987). *Sāṁkhya: A dualist tradition in Indian philosophy*, Vol. 4 of Encyclopedia of Indian philosophies. Delhi: Motilal Banarsidass Publishers.
Long, J. D. (2011). Jaina Philosophy. In J. L. Garfield & W. Edelglass (Eds.), *The Oxford Handbook of World Philosophy* Oxford: Oxford University Press.
Matilal, B. K. (2002). *Ethics and epics: The collected essays of Bimal Krishna Matilal* (Vol. 2). New Delhi: Oxford University Press.
Mohanty, J. N. (2000). Moral philosophy. In *Classical Indian Philosophy*. Maryland: Rowman & Littlefield Publishers Inc.
Pániker, A. (2010). *Jainism: History, society, philosophy, and practice*. New Delhi: Motilal Banarsidass Publishers.
Roach, C. (1996). Loving your mother: On the woman–nature relation. In K. J. Warren (Ed.), *Ecological feminist philosophies* (pp. 52–65). Bloomington: Indiana University Press.
Rolston, H, I. I. I. (1987). Can the east help the west to value nature? *Philosophy East and West, 37* (2), 172–190.
Sasidhar, K. (2006). Ecology and Vedic vision. In R. Balasubramanian (Ed.), *The enworlded subjectivity: Its three worlds and beyond, philosophy and culture in Indian civilization* (Vol. 11, pp. 697–724) (Series editor. D. P. Chattopadhayaya). New Delhi: PHISPC, Centre for Studies in Civilizations.
Soper, K. (1995). *What is nature? Culture, politics and the non-human*. Oxford: Blackwell Publishers Ltd.
Sovani, V. V. (2005). *A critical study of the Sāṅkya system*. Delhi: Chaukambha Sanskrit Pratishthan.
Taylor, P. W. (1986). *Respect for nature: A theory of environmental ethics*. Princeton, NJ: Princeton University Press.
Warrier, A. G. (1983). Krishna (trans.). In *Bhagavadgīta, SrimadBhagavadgīta Bhasya of Sri Śaṇkaracharya*. Madras: Sri Ramakrishna Math.
Wiley, K. L. (2006). Nature of nature: Jain perspectives on the Natural world. In *Jainism and ecology: Nonviolence in the web of life*. New Delhi: Motilal Banarasidass Publishers.

Chapter 6
Topocentric Views of Nature

Abstract Drawing from different secular traditions of India, such as the Ayurvedic health traditions and Sangam literature, geographical description and land ethic practices from South India, this chapter describes place-centric views of landscape which look at nature as a habitat for human beings. While one tradition relies on typology of the human body and its relationship to geography and climate, another poetic tradition from South India utilises topocentric categories called *tiṇai*. Earlier work by scholars in these areas is summarised with appropriate annotation and analysis that links these ideas to the topocentric view of nature.

Keywords *Tiṇai* · Nature as habitat · Topocentric view · *Āranya* · Forest · *Āyurveda* · Wetland · Dryland · Jaina geography

6.1 Introduction

If one was to close one's eyes and try and visually imagine nature, what pictures would form in one's mind? It is most likely that the picture would be of some scene of a landscape, maybe of mountains or rivers or forests. Given that the modern usage of the terms nature, landscape, and scenery are similar, it is also important to study some ideas of landscapes in other textual traditions in India. Brunn and Kalland (1995, p. 10) suggest that the interpretations of environmental issues are anthropocentric in the sense that human beings subjectively seek order in the universe through the conceptualisation of nature as landscape: "We differentiate nature by means of accurate definitions, classes and systems, including shapes and colours but, attribute to it meanings and emotions turning the environment into landscapes".

Though landscape is a constructed category, still it has come to represent nature for us in many ways. Baindur (2010c) suggests:

> From a phenomenological perspective, one could conceptualise nature based on the analysis of the human experience of the world, understanding it as a discourse, and not as

M. Baindur, *Nature in Indian Philosophy and Cultural Traditions*,
Sophia Studies in Cross-cultural Philosophy of Traditions and Cultures 12,
DOI 10.1007/978-81-322-2358-0_6

mere physical reality. Such an interpretation would explore the way in which one makes sense of the phenomena of the surrounding world and the ways in which a human being is understood to be intertwined in nature.

One might argue that these views do not strictly fall under the philosophical understanding of nature, but given the history of Western thought, it is clear that the Romantic Movement's perception of nature challenged the mechanistic concept of nature. These gave rise to some of the very important debates around the relationship of human beings to nature. It is within the same theme that I seek to understand the categories of landscape features that for the human imagination, are the closest way to identify nature.

There are alternate ways to look at the cultural links between geographical categories and nature. Unlike the geographies of sacred landscape that greatly rely on religious mythology and traditions (to be discussed separately in Chap. 7 of this book), these conceptualisations are closer to our modern understanding of nature as landscape. We will engage with some of the topocentric views of nature from two very different traditions of thought that are not based on philosophical sources and are connected to two non-religious traditions of thought: medicine and geographical literature. The first is a summary of a topocentric view of Āyurvedic categories. The book, *Jungle and the Aroma of Meats,* by Zimmerman (1999) is a good starting point to explore this area of medicine-based topography. This will be followed by an explanation of some categories that are equivalent to wilderness and human settlements, and the relations between forests and people in India where one can bring together ideas from historians and environmentalists. The second topocentric view is from the concepts of Tamil literary traditions that combine topos (landscape) and human dispositions through a unique category of landscape descriptors called *tiṇai.* Though scholars would prefer to treat this as a literature project, more suitable to eco-criticism themes, I prefer to classify *tiṇai* under a topocentric theme. The attempts to link landscape and human emotion are to be found in almost all kinds of poetry and composition, but the analysis and acceptance of this as a literary theory in the poetics has been prevalent much earlier than the discipline of eco-criticism itself that comes to us from the West. In a later chapter, I will use the methods of eco-criticism to analyse other pre-modern Sanskrit literature. The sensibilities of Tamil literature are also different from the rather direct references to nature in other compositions. We will briefly examine the idea of Jaina geography through one of their classical texts. To add to these rich interpretations of landscape, a history of the Bishnoi people's relationship to their land will be described.

6.2 On a Topocentric Ecology and Health Traditions

Callicot (1987) points out that there was a popular view of Eastern thought in the West and many people who read and studied some of the Eastern philosophies had a profound impact on certain sections of Western thinkers. This in turn led to the development of a very general and popular view of all Eastern philosophy which he rejects as the "shallow view of Eastern thought". He points out that, conceptual resources for environmental ethics in Eastern traditions of thought require further research: "… there is an opportunity for students of Eastern thought to contribute in a most welcome and important way to the literature of environmental ethics" (p. 47).

Brennan (2002) contests Callicott's position and argues that in the general survey of ecological thought among Hindu traditions, particularly *Āyurveda*, one of the traditional systems of medicine in India has been ignored. And he claims that in this system of medicine is embedded an ecological understanding that is related to the people's life within an environment. This is reflected in the texts that describe the interrelationship between people and places. He explains:

> The quality of a place that emerges seems to be predominantly anthropocentric: an ecology of agriculture cooking and pharmacy, not in an ecology of interacting systems defined without a reference to human meanings and practices. It is a profoundly human ecology, in other words, yet one which sees the world in terms of processes linking different individuals in populations (within and across species)—for example, the processes of concentrating essences through feeding (Brennan 2002, p. 574).

In his paper, he concludes that it was important to pay attention to the "place-centric topographical medicine". He writes: "It is this crucial move, the 'topo-centric' focus of the theories which provides a kind of ecological understanding of the human subject, of the subject in relation to the surrounding world and that of the world itself" (p. 580).

He strongly argues that to dismiss this view of ecological health as anti-environmental is unjustified and that a conclusion based on traditional conceptions of body and place is inconsistent with modern understanding would be a hasty decision, because scientific understandings may not be the only legitimate interpretations of the world around us.

Brennan (2002) therefore recommends a study of the way *Āyurveda* interprets the surrounding and its relationship to the human being from an ecological perspective. He further adds that the system of medicine is closely intertwined with agricultural, seasonal, and food practices; hence, it has implications for ecological health, the valuing, and managing of nature, as well as for biodiversity. An interesting view of landscape in Indian thought comes from the tradition of *Āyurveda* as described by Zimmermann (1999). It gives us (something) to understand how the idea of the jungle and topocentric views seems to be inclusive of the human beings in an essential way. In his book (that Brennan refers to), he uses the perspective of *Āyurveda* to understand some topocentric views in Indian thought. The categorisation of landscape as wet (*ānūpa*) and dry (*jāṅgala*) and the relationship of human–Ayurvedic body types to these landscapes are discussed in detail. Zimmermann suggests a whole new perspective to the study of classical Indian medicine. In the foreword to the book, Wujastyk (1999, p. xv) remarks that this work brings into the field a whole new creative and "interpretative sophistication" to the field of medical ethnosciences. Zimmermann highlights the unique relationship between human beings and their habitats within the context of health and environment. Some of the main concepts in this book are firstly, the idea of *cara*, the human interaction within food and environment. Secondly, the interconnection between the human body type and the land typology is linked to the presupposition that the microcosm and the macrocosm are often reflections of each other. Earlier traditions of Indian thought also perceive creation as a cosmic person and the human being as the universe. These kinds of presuppositions could also be interpreted as symbolic rather than actual. One could say that such traditions perceived the universe through a person-centric framework. The common ground of experience between creation as nature and the human subject is the functional body of the individual.

In later Āyurvedic texts, one finds that an interpretation is not so much about the similarity of the body of nature and the body of the human being, but the relationship between the two based on compatibility and non-compatibility.

The human body in Āyurveda is seen as a balance of body constituents (*doṣa*) called *vāta, pitta*, and *kapha*. When they are in balance, they create combinatory body constitution of most individuals that have two predominant constituents such as *vāta–pitta* or *vāta–kapha*. Imbalance of these constituents by agitation of any of these three *vāta, pitta*, and *kapha* causes illness. The *doṣa*s are activated beyond normal differently during different times of the day, during different seasons, during different stages of one's lifespan, or during times of mental or physical afflictions. The internal environment given by constitution of these interacts with the external environment given by the wet, dry, and middling landscape, producing effects of either adaptation called *sāmyatā*—harmonious state—or a state of agitation, of the *doṣa*s.

The similarity of human body types *vāta, pitta*, and *kapha* to a classification of topology and landscape as (windy) dry, middling, and wet, described in āyurvedic texts, is about relating the inside to the outside. Zimmerman's study linking these types to the actual geography of India as wetlands—*ānūpa* and dry lands—*jāṅgala* indicates that the medicinal system of Āyurveda may have laid emphasis on the interactions between human health and nature at a much deeper level than is popularly known today. While lifestyle is often stressed in today's prescriptions of Āyurveda, Zimmerman's work seems to suggest the linkages between food, body type, and propensity for illnesses and the geographical landscapes. The aetiology of the system can be critiqued in the light of modern developments of both *Āyurveda* and medical sciences, but it is clear that the conceptualisation of the climate in the natural landscape as similar to the climate (body type) landscape of the human bodies is an important ecological theme. While *jāṅgala,* or the dry and windy lands, are good for habitation, it is likely that a person with a *vāta* complaint will find the wet locale more suitable. Zimmerman (1999, p. 21) demonstrates that in *Āyurveda,* nature and the human being are linked through interactive relationships of geographical inhabitation. To inhabit is to create ways of routine adaptive behaviour to the environment which is captured by a Sanskrit word, *cara*. In some sense, geographical understanding of nature as place is categorised by a human perspective of health and well-being, in other words a sort of biological category of habitat. On the other hand, this understanding is not merely about landscape but also about the various roles and functions of human beings—their food, their activities, their life cycles, and their interaction with the landscape.

Zimmermann (1999) places importance on the human interaction with the environment through the concept of *cara*. He remarks:

> *Cara* is perfect example of a word with a double reference: to nature and to man. The objective or spatial reference—the environment—is incorporated within the subjective or practical reference—the environment regarded as a source of the means of subsistence (p. 21).

The word *cara* in Sanskrit refers to moving or roaming. It also refers to behaviour and eating. Based on a presupposition that one absorbs the essences of the soil called *rasa* through food, it is often concluded in many texts that a human being is subject to influences of the environment and habitat of the origins of food that is consumed. Traditionally too, we find in popular beliefs in practice that rice from the lowlands and from the dry lands are supposed to have different qualities. Drawing from Zimmermann's arguments, one could say that there is another way that the environment affects human beings internally. The essence (*rasa*) of the soil gets absorbed into the body of the eaters through the plants and the flesh of animals. Food, thus, becomes threefold in its function. Primarily it nourishes the body in a homologous sense. Fluids augment fluids of the body and so on. This is the general feature of the nutrients drawn from food. As a prescription, the same food that contains the essence of the soil becomes medicine in a particular quantity and form. Yet again the same food becomes poisonous when the quality and quantity is not right or when the time and season of consumption is disharmonious.

Diet, food, and activities are thus prescribed according to seasons and the time of the day based on a concept called *ṛtucarya—cara* according to seasons. Repetitive seasonal practices and routine include various prescriptions of different types of food to be cooked and eaten based on one's activity, gender, and also age. While Ayurvedic prescriptions are based on classical texts, the practice of *ṛtucarya* is more traditionally practiced at community and household levels. Texts on *ṛtucarya* do contain guidelines and are non-classical in the sense that they do contain reference to non-vegetarian and local diets. This suggests that these texts were not ritual based or limited to study by the elite castes. It must also be pointed that the seasonal practices related to land and food were implemented at the household level by the couple, the woman usually keeping track of *ṛtu* (seasons) using the festival days to mark seasonal changes in diet and seasonal practices. I recollect my grandmother announcing that we had to pack away our warm clothes after the festival *Śivarātri* (a festival for the god Śiva) in February–March, after which the skin had to be gently exposed to the warming influence of the sun to acclimatise us to the oncoming spring and summer. There may not be an authoritative reference to this kind of prescriptions, but they do exist among womenfolk and among many traditional communities. They establish a strong tradition of topocentric and season-centric relationship to the environment. Zimmermann's interpretation of the Ayurvedic sciences as embedded in the geography of wet and dry lands gives us an understanding of nature as a habitat that is implicit in a medicinal tradition.

6.3 The Geography of the Forest

The forest, or the *araṇya*, is one of the geographical categories of landscape that is popular in almost all kinds of literature including mythologies and epics. As a habitat, the forest is the dwelling place of the non-human. Besides the wild animals (*mṛga*), the forest is also home to ghosts and ghouls, to forest spirits and water nymphs. While one can, for purposes of empirical validity, ignore references to these mythical beings, it is possible to infer from these textual references that forests are not forbidden for human entry but they certainly are not the normal dwelling places of human beings. In a verse in *RgVeda* (*RV* X.46, trans. Panikkar 2001), "*Araṇyānī*" (wife of the forest), the poet of the *Veda*, addresses the spirit of wilderness or the goddess of the forest. One of the earliest references to the idea of a nature perceived as wilderness versus a culture that is seen a habitation of the villager can be seen in this verse. The comparison between the settlement and the uninhabited forest is clearly marked out in the many images of this poem: "… how is it you avoid people's dwellings?" and further: "She needs not to toil for her food, mother of untamed forest beasts". The idea of the polarity between *āraṇya* and *grāmya* is well described by scholars of history such as Zimmerman (1999, p. 101): "Throughout Sanskrit literature *grāmya* (domesticated) is opposed to *āraṇya* (wild/belonging to the forest)". Thapar (2012) also points out that "the dichotomy between the *vana* and the *grāma* evolved in early time when the village constituted the settlement". The idea of the *āranya,* however, is different from the category of a pristine nature. A continuous interaction of human beings with the forest is not unusual. As illustrated in the verse, again we see woodcutters and the cowherd moving through the forest landscape (*RV* X.46, trans. Panikkar 2001) and the benevolent forest offering up her delicious fruits for consumption when approached with non-violence. The terms *āraṇya* and *grāmya* are dichotomous and inclusive "and perceptions accompanying it were neither static nor uniform" (Thapar 2012, p. 106). She claims that these terms capture for us the symbolic value of the landscape. I suggest here that the binary of the *āraṇya* and *grāmya* ignores the interspaces of topography that are referred to in texts and common usage. Notions of the idea of wilderness conceptually influence the modern perception of *āraṇya* as a place for the non-human.

In a short project that I conducted on concepts of landscape in a small village close to Bangalore, villagers were interviewed about their categorisation of land. Out of this survey, category distinctions were made by these 20-odd villagers between arable land (farms), pasture land, village, and the market (the city was a market). The interspaces between a forest and a settlement are occupied by a series of changing topos, not only divided spatially but also temporally. These interspaces become the places where the human and the non-human encounter each other. Each of these places can be adjacent to or even enclosed by another functional type of landscape. The landscape type is also connected to the available water source and natural resources that are present in the place.

On the other hand, if we look at the concept of *āraṇya*, *vana*, and other cultural terms describing landscape, they are used in specific functional contexts in literature.

We find that they are not used so synonymously. An *āraṇya* is a deep, dark forest almost equivalent to wilderness. Humans do not dwell here, unless they are sages. Rāma kills the wicked demoness Tātakā in *Daṇḍakāranya*. The *vana*, on the other hand, usually denotes a less dense category, often like a grove of trees. The *vana* is the forest for exile, the "out of settlement" region; exile is called *vanavāsa*. Sacred groves are also designated as *divyavana*. The edges of a forest that are close to the settlements were called *upavana*. The *āśrama* in the forest is a human settlement within a forest as much as an *upavana* is a wooded area within a settlement. When the rural *grāmya* becomes the town, *nagara,* the wooded area is included within the human settlement, as *udyānavana.*

Many of the Upaniṣads were themselves created in the *āraṇya*, sometimes called *āranyaka.* The forest thus had its mystical influence on the path of liberation through a symbolic act of leaving the *grāmya* (village) for the *āranya* (forest). Thapar (2012, p. 112) suggests that this act emphasises the solitary journey of an individual for purposes of asceticism and has many interpretations such as "... distancing from civilization, seeking of knowledge through isolation and meditation; and a search for the meaning of life through experiencing the unknown". Since the social obligations were reduced, there were fewer intrusions into the practices of the aspirant who wished to be liberated from rebirth (Thapar 2012, p. 112). The very context of the forest then forms a backdrop to the creation of ideas that were inspired by the solitude and wilderness—the Upaniṣads.

6.4 *Tiṇai*: Landscape and Poetry

Tiṇai refers to the bio-geographical regions mentioned in the *Caṅkam* Tamil poetry works. These are categories of natural regions that are described in detail in the *Tolkāppiyam*, a grammatical work of much importance on the poetry and grammar of *Caṅkam* literature (Sivathamby 1974). The literary and cultural movement during the *Caṅkam* period in the south of India gave rise to a number of poetical works which were based on compositions created during three assemblies held in the town of Madurai, the capital of the *Pandya* kings (Murugan 2008). These works form a significant landmark in the classical literature of the Tamil people. The culture and language of the original inhabitants of the south of India is referred to as Dravidian as against the Aryan culture that was predominant in the north. The Dravidian culture was based on a classical form of Tamil, one of the few languages in India that is not Sanskrit based. Zvelebil (1992) dates the period of composition of these works from 200 BC to 200 AD. The literature of this era spans a rich tradition of poetry and commentary including the *Tolkāppiyam* (henceforth, *TKP*), a grammatical work on these poems composed by the author referred to as *Tolkāppiyanar*.

Scholars who work on *Caṅkam* literature use the term called *turai* to describe to the literary categories that refer to "context" or the twofold classification of poetry called *agam* and *puram*. While *agam* poetry deals with themes related to family and love, *puram* deals with affairs of court, military conquests, and heroes. Zvelebil (1992) refers to this classification thus: "In classical Tamil Poetry—in its *pre-bhakti*, *pre-Pallava* age—a self-conscious, indigenous Tamil culture is depicted in its two principle forms, the interior (lyrical *agam* hyper genre) and the exterior (lyrical *puram* hyper genre)". Of the two, it is the *tiṇai* of the *agam* poetry that conceptualises nature in its complexity, intertwined with the human beings and their activities. The poetry besides describing nature has a deeper layer of meaning.

There are seven *tiṇai*s in the *agam* context, each representing the poetical mood of love. Five of these can be said to be "natural" relationships having physiographical content or natural "well-matched" love situations, and they are named after a plant endemic to that region. Two categories of *tiṇai*s describe "unnatural" moods of love, *peruntiṇai,* and *kaikilai,* which were in contrast "ill-matched". Gurukkal (2010) calls these categories as "ecotypes" or geographical descriptions of place. He analyses the eco-semiotic view of *tiṇai* in the context of Tamil poetry. He points out that the *tiṇai* are ecological signs and have a metonymic structure. These signs, according to him, have evolved from the ecosystems and attained what he refers to as "… cultural signification through linguistic and aesthetic practices of peoples adapted to these different ecosystems…" (p. 78).

Sivathamby (1974, p. 25) on the other hand, explains that the five *tiṇais* were contemporary physical realities[1]. He also argues that the physiographic division of the Tamil landscape is actually four in number, excluding the *pālai* (desert), indicated by the fact that the Tamil landscape is often called *nānnilam* (land that has four types). With evidence from literature, he demonstrates that the desert landscape *pālai* arises from the other four *tiṇais* during particular seasons or during drought. He then suggests the five "landscapes," (*tiṇais*) into which space is organised, that correspond to the major ecotypes (*nilam*) of the Tamil region. These are (from Selvomony 2008, p. 25–26):

1. Hill, *Kuṛiñci* (*Phelophyllum kunthaianum*), named after a flower predominant in hilly tracts.
2. Field, *marudam* (*Terminalia* sp.), which grows where the soil is alluvial with ponds, water buffalo, water birds, and fish.
3. Pasture, *mullai* (*Jasminium* sp.), named after a type of jasmine flower.
4. Seashore, *neytal* (*Nymphaea* sp.), after a water lily which is a characteristic flower of the region.
5. Wasteland, *pālai*: desert land, arid region with the shrub *Wrightia tinctoria.*

Before describing these landscapes in detail, it is important to also examine the classification of the components of *tiṇai.* According to the classical analysis of *Caṅkam* poetry, *tiṇai* is co-constituted by three elements (Takahashi 1995; Dubyanski 2010). The classical commentary *Tolkappiyam Porulatikaram*

[1]Sivathamby (1974), in his paper on the social origins of *tiṇai* concept, also points out the historical, cultural, and social significance of these categories. He attempts to examine uneven patterns of development and social organisation and analyses them.

(*On Tolkappiyam,* hence forth *TP*), describesthat *tiṇai* is co-constituted by three elements—(1) primary or (*mutal*), which consists of basic things such as tract of land (*nilam*) and time (*poḷutu*), (2) germinal elements or things that are born (*Karu*) consisting of things that grow or that transform in the environment such as god, food, beast, flower, bird, occupation, tree, drum, and musical instrument, and (3) the specific *(uri)*: feelings, behaviour, and situation (Takahashi 1995). Each *tiṇai* is related to geographical elements, a season, a time of the day, occupations, and the moods of human beings in relationships with each other (Baindur 2010c). What is interesting about this conceptualisation of nature is that it includes the human being within its fold and is described as a poetic background of nature for human activity.

For example, in the poem representative of the seashore, *neytal tiṇai* (extract from Kuruntokai 325, trans. quoted from Selby 2008, pp. 21–22):

What she said:
...
O Mother,
our master who supports us—
where is he now, I wonder?
The place between my breasts
has filled up with tears,
has become a deep pond
where a black-legged
white heron feeds.

We find that the sea is represented by the salty tears of the heroine. The pond and marsh all reflect the mood of lament of a lover's absence. According to Selby (2008), the very body of the heroine becomes the landscape, each reflecting the situation of the other. While the mistrust of the heron is similar to her lover, she herself feels as if she has been the fish. Tuan (1974, p. 93) describes the relationship between landscape and human beings as "topophilia". According to him, this word describes "all the affective ties of a human being with the environment". It is only natural that the aesthetic response of the *Caṅkam* age to nature should find their expression in poetical aesthetics. Moreover, it is clear such a conceptualisation of nature that includes the human being can be attempted only through the medium of a language and through categories of interpretation that express this relationship in all its complexity, that is poetry (Baindur 2010c).

6.5 *Tiṇai* as Nature-Place or Nature-Landscape

Are the *tiṇai* mere literary categories or are they related to the real landscapes? If they were mere poetic categories, then the imagination of *tiṇai* as nature is limited only to literature and its engagements and would not be as relevant for the cause of ecological thinking. I posit that the *tiṇai* describes the metaphysics of nature as inclusive of the human.

Dubyanski (2010) points out that in *TKP*, the author refers to the five regions that are called "parts of the earth" (Sutra III, 5; 951 in V. Murugan's edition, quoted by Dubyanski 2010):

> The world of forests where *māyoṉ* (Naryana) dwells,
> The world of dark mountains where the Red-one (Murugan) dwells,
> The world of sweet waters where *Indra* dwells,
> The world of spacious sea sands where *Varuna* dwells.

He also suggests that according to *TKP,* there is a definite correspondence between human situations and natural background. Three poetical themes that do not correspond to the actual geography are excluded in this description—*pālai:* desert land (as a degraded state of other types) and the two ill-matched *peruntiṇai,* and *kaikilai.* This could create an objection to the argument that *tiṇai* are mere poetical categories that have no significance to the natural world. Selby (2008) explains that while *tiṇai* is the artistic space in poetry, it is based on actual ecotypes called *nilam.* However, this seems to be based on the conceptualisation of nature as non-human. Even if we consider non-human components of *tiṇai, nilam* or land is only one constituent of "place"; the other constituent *karu*, which includes flora and fauna, is equally a part of the natural world.

Perhaps one could suggest that though *tiṇai* is a conceptualisation of nature in poetry, it is still based on real geographical elements of human experience. We cannot, however, relate direct abstract categories of meaning-making in poetry to the phenomenon of experience. However, given that *tiṇai*s are named after eco-system signifiers, it is possible to understand it as poetry being derived from geographical experience.

Murugan (2008, p. 11) also emphasises the deeper connection between the natural and the human by pointing out the intertwining of these elements in the complex of *tiṇai*. He writes:

> For here, nature is not conceived as simply a backdrop to the human drama as is the case with most poetry of the world. It is not even a mere evocative background to the play of human emotions and deeds. The rhythms of human life and those of nature are made to correspond, coexist, and coalesce in these poems.

The conceptualisation of *tiṇai* entails a deep connection of the land and its belonging. Belongingness expresses as belonging to a place, where the sense is given by natural landscape (Baindur 2010c). According to Tuan (1974, p. 132), early civilisations historically show the transformation of the idea of a cosmos into landscape. He calls this "axial transformation". Landscapes according to Tuan (1974, p. 141), served the purpose of being "the background for commonplace human activities". He posits that the world of the pre-modern man was rich in symbols and metaphors, which evoked emotional responses to nature that was multi-layered and ambiguous.

Andrew and Duncan (1989, quoted from Anderson 2010, p. 39) describe place as constitutive of three parts: location, locale, and sense of place. I suggest here that location indicates an objective point in space represented by a grid or axis of reference, locale describes the background of natural, social, and built environment that makes every day human experiences possible. It is often the feel and ethos of a place. The third part of place is the "sense of place" and it relates to the affective component of human beings and space. Comparatively, it seems that the three elements of *mutal* (time and place) are similar to "location", while *karu* given by the elements describes a sense of the "locale", and finally, *uri* is the specific situation representative of the "sense of place". In fact, it is the human emotion that finally designates the *tiṇai,* explains Selby (2008):

> In fact, the *Tolkāppiyam* stresses that emotion (or mood) is the only thing within a *tinai* that is actually fixed, a rather difficult concept to grasp, but crucial to the understanding of this system. *Akattiṇaiyiyal* verse 13 states: "The things that are not behavioral elements may overlap," meaning that everything except for the behavioral elements may (p. 25).

6.6 Jaina Geography

The Jaina texts are very descriptive of the geography of landscape or the middle worlds that human beings inhabit with all other kinds of beings. These are also somewhat secular descriptions of landscapes of the great continent called *Jambhūdvīpa* (island of the rose apple/Indian blackberry tree) which are highly imaginative. These descriptions are not story narratives like sacred landscapes stories of sites of Hindu pilgrimage centres. In that sense, these descriptions are secular. But in another sense, these narratives form a part of the soteriological concerns of the philosophy—the idea of *karma* and liberation for the Jaina thinkers. This cosmic geography is described in some of the primary texts like the *Jainā-gamas* and also in smaller texts like the *Jambhūdvīpasaṃgrahaṇī* (henceforth, *JDSH*) of Haribadra Suri composed in *Mahārāṣṭri*[2]. The elements of the human and non-human world are organised into geography and cosmography through detailed description of directions, sizes, and areas of the cosmos. Pániker (2010) writes that the geographical concerns of Jainas were connected to their requirement to cate-gorise the various beings of their cosmos in their designated habitat worlds: "The description of the physical geography of *Jambhūdvīpa*, as detailed as it may appear in the text and illustrations, is not the most relevant consideration, what is funda-mental, let us stress again is the moral and soteriological geography" (p. 42).

According to the Jaina texts, only two and a half spheres of the seven continents on *Jambhūdvīpa* are subject to moral degradation and the effects of *karma* and its fruits (p. 42). The rest of the cosmos remains constant in virtue, body size, climate, or vegetation. The part of the cosmos that undergoes change and the suffering of *karma* is called *karmabhūmi*:

[2]A form of Prakrit. The version referred to for this book is with commentary in Sanskrit by Prabhananda Suri, critically edited and translated by Frank Van Den Bossche.

> It is in these regions that human beings have to work and act in order to get by, and consequently it the where the law of *karma* imposes retribution according to action. Fortunately, it is precisely in these regions that humans can have recourse to asceticism to be able to 'burn' *karma*, attain enlightenment and final liberation (p. 42).

Bhāratavarśa is described as the southern continent separated from the rest of *Jambhūdvīpa* by the Himalayas. Only a part of this region, the south central part, is called *Āryakhanda*, where the reach of Jainism prevails. This is the region where one may aspire for liberation or a higher birth in one of the other *karmabhūmi*s that are more conducive for liberation, where the chronological period ensures presence of a *tīrthaṅkara* (an enlightened soul who is a guide for liberation) and certainty for human liberation.

The *JDSH* (Haribhadrasūri) in particular describes the various regions, mountains, and rivers, naming them with certainty, with a number of references to distance and directions in the forms of *sūtra*s with a commentary. The text sets out its agenda in the second *sūtra*:

> 1) sectors, 2) the *yojana*s, 3) continents, 4) world mountain ranges, 5) peaks, 6) fortresses, and the rows (of abodes), (7?) 8) provinces, 9) the mountain lakes, and 10) rivers The sum total of these [constitute] the *samgrahiṇi* (summary) (Haribhadrasūri, *JDSH Sūt*. 2, p. 41).

One of the most interesting aspects of Jaina geography in the *JDSH* is the listing of numerous names of places and regions. Apart from naming so many regions, the geography is mathematically mapped out according to the canonical descriptions of earlier masters and teachers. We can see from the *Sūtra* 2 described earlier that the Jaina world is divided into natural features that are somewhat similar to our own modern descriptions of landscapes. The mountains are called *giri* and *parvata*, and they have sharp peak features called *pavatakūta*. The abodes in a row are those of divine beings often constructed with precious metals and ornate pavilions and gardens. The *kṣetra* or *varśa* refers to the continent and also the great *Jambhūdvīpa* in general. The text is elaborate in its descriptions with numerical distances and sizes that claim the area of regions and heights of the peak. For instance, in the sixth and seventh *sūtra*, the calculation of the area of a wall with a lattice and a balcony around *Jambhūdvīpa* which is circular is mathematically explained. The commentator acknowledges Aryabhatta I's method of calculation is to be followed: "The circumference of a circle is the square root of ten times the square of the diameter. Its surface area is the circumference multiplied with a quarter of the diameter" (Haribhadrasūri, *JDSH Sūt*. 2, p. 64).

This text seems to suggest that the Jaina philosophers were very interested in naming and describing the geography of the world they lived in. Yet, the logic of calculation through mathematics informed them of the validity of their conclusions. After explaining that there are 11 peaks in *Himavat* and *Śikharin,* the author derives 61 peaks on the mountains and arrives at a total sum of 467 peaks in all (Haribhadrasūri, *JDSH Sūtra* 15, p. 178). The number of peaks in each of the listed 61 mountains is multiplied by various numbers of peaks each to give the total number. Two sets of numbers are given in relative order to be multiplied by another series in order to give the final total: relative orders being {11,2,52,2} numbers and {4,7,9 and 11} peaks (Haribhadrasūri, *JDSH Sūtra* 16 p. 181).

For instance, the *Himavat* and *Śikharin* have 11 peaks (*kūtas*) and are to be multiplied by 2 to give a total of 22 peaks. The text is difficult to follow without the commentary that clarifies a great deal of these mathematical operations that arrive at different figures and numbers of measurement and counts of landscape features.

The Jaina geographical imagination seems to be an exercise in mathematical imagination and abstraction of landscape itself. It may contain some geographical information in some places, but one could almost call this a mathematically imagined geography. The description of operations of numbers and the various calculations of area seem to be significant for the author than describing actual landscapes, climates, flora, and fauna. Small descriptions of some of the landscapes are present in the commentaries that are not in the main *sūtra*. For instance, "*Mahādrahah*" refers to "great lakes", or mighty bodies of water, much larger than other lakes (*Sūtra* 20, commentary, p. 204). But these are insignificant compared to the numerical descriptions of rivers—how wide, how deep, and how long, mountains—how high and the circumferences at the base and on top, or continents—how large an area. *Sūtra* 27 on the mountain descriptions, for instance, reads thus: *Śikharin* and *Kṣullmahimavat* are one hundred *yojanas*[3] high and are made of gold. *Rukmin* and *Mahāhimavata* are two hundred [*yojanas*] high and made of silver and gold (p. 242). Further describing the mountains, the *sūtra* claims that they are rooted under the surface of the earth, one-fourth part of their height (*Sūtra* 28, p. 254).

In conclusion, one may say that the precision involved in describing the geography is to delineate the regions of *karmabhūmi* with accuracy. It is also to indicate the places where liberation is possible and where one could find the true teachers of Jainism, the *tīrthankaras*. In other words, where one is located, in Jaina thought is to be articulated in detail to let one know what one has to do to advance in the spiritual path. This knowledge would possibly motivate one to seek the lands of higher possibility of liberation.

References

Anderson, J. (2010). *Understanding cultural geography: Places and traces*. Oxon: Routledge.

Baindur, M. (2010c). *Nature as being: Poetry and place in the Tiṇai complex*. Paper presented at the national conference on 'The Name and Nature of Ecocriticism,' organised by Department of English, School of Social Sciences and Humanities in collaboration with Organisation for Studies in Literature and Environment-India (www.osle-india.org) on 6 and 7 December.

Brennan, A. (2002). Asian traditions of knowledge: The disputed questions of science, nature and ecology. *Studies on the History and Philosophy of Biological and Biomedical Sciences, 33*, 567–581.

Bruun, O., & Kalland, A. (Eds.). 1995. Images of nature: An introduction. In *Asian perceptions of nature: A critical approach* (pp. 1–24). Richmond: Nordic Institute of Asia Studies, Curzon Press.

[3]One *yojana* is roughly 15 km (from Pániker 2010).

Callicott, J. B. (1987). Conceptual resources for environmental ethics in Asian traditions of thought: A propaedeutic. *Philosophy East and West, 37*(2), 115–130.

Dubyanski, A. M. (2010). 'Tolkappiam.'*World classical tamil conference souvenir*. Kovai, pp. 72–79.

Gurukkal, R. (2010). Semiotics of ancient Tamil poetics: A methodological consideration. In *Social formations of early South India* (pp. 77–91). New Delhi: Oxford University Press.

Haribhadrasūri. Bossche, F.V.D (Trans.). (2007). Elements of Jaina geography: *The Jambūdvīpasaṃgrahaṇī of Haribhadra Sūri: critically edited and translated with the commentary of Prabhānanda Sūri*. New Delhi: Motilal Banarsidass Publishers.

Murugan, V. (2008). Cankam literature in Tamil. '*Pudiya panuval: An international journal of Tamil studies, North America*.1.1. (online). http://www.indianfolklore.org/journals/index.php/Panu/issue/view/52/showToc. Accessed on July 22 2010.

Panikkar, R. (2001). *The Vedic experience: Mantramanjari* (1st ed.). 1983 reprint. New Delhi: Motilal Banarsidass Publishers.

Pániker, A. (2010). *Jainism: History, society, philosophy, and practice*. New Delhi: Motilal Banarsidass Publishers.

Selby, M. A. (2008). Dialogues of space, desire, and gender in Tamil Cankam Poetry. In M. A Selby & I. V. Peterson (Eds.), *Tamil geographies: Cultural constructions of space and place in South India* (pp. 17–42). State University of New York Press: Albany.

Selvomony, N. (2008). tiNai in primal and stratified societies. *Indian Journal of Ecocriticism, 1*, 9–21.

Sivathamby, K. (1974). Early South Indian society and economy: The Tinai concept. *Social Scientist, 3*(5), 20–37.

Takahashi, T. (1995). *Tamil love poetry and poetics*. Leiden: Brill.

Thapar, R. (2012). Perceiving the forest: Early India. In M. Rangarajan & K. Sivaramakrishnan (Eds.), *India's environmental history—A reader: (Vol. 1: From ancient times to the colonial period* (pp.105–126). New Delhi: Orient Black swan.

Tuan, Y. F. (1974). *Topophilia: a study of environmental perceptions, attitudes and values*. New Jersey: Prentice-Hall Inc.

Wujastyk, D. (1999). 'Foreword.' In Zimmerman, F (Ed.), *The jungle and the aroma of meats: An ecological theme in Hindu medicine*. New Delhi: Motilal Banarsidass Publishers.

Zimmerman, F. (1999). *The jungle and the aroma of meats: An ecological theme in Hindu medicine*. New Delhi: Motilal Banarsidass Publishers.

Zvelebil, K. (1992). *Companion studies to the history of Tamil literature*. Leiden: E.J Brill.

Chapter 7
Sacred Geographies and an Ethics of Relating with Reverence

Abstract After secular ideas of landscape ecology, this chapter deconstructs the understanding of sacred nature and natural landscapes in the subcontinent. Religious and philosophical thought has always influenced people's social and ecological behaviour in India. The unique world views of nature in Indian thought through relationships between place, the idea of sacred, and narratives about sacred landscapes called *sthala purāna* are elucidated. The chapter also explains how secondary narratives called *sthala māhātmya* recount the human experience of the sacred and create a moral relationship between landscapes and people. As practices around sacred geography and pilgrimages are prevalent even today, I conclude this chapter with suggestions of the possible place centric, relationship-based ethics of sacred landscapes.

Keywords Sacred landscapes · Narratives · *Sthala purāna* · Place · *Sthala* · *Sthāna* · Sacred geography · Reverence to nature

7.1 Introduction

Nature in India has been conceptualised as sacred, both within the mainstream Hindu religious tradition as well as within other belief systems of communities, including indigenous communities such as the *Adivāsi*s. The presupposition that nature and natural objects have sacred meanings for such communities can be inferred from the practices of reverence of nature such as worship or other rituals and popular narratives involving nature as sacred. The reverence for nature within such practices is embedded in the everyday lives of people, overtly expressed as some form of worship, ceremony, or daily practice such as drawing patterns of *rangoli/kolam* on the ground, tree, or water worship, and demarcation of sacred groves. It would not be wrong to say that conceptualisation of nature within the Indian world view is constructed to a great extent by narratives, myths, and ritual. This according to Nagarajan (2000a) can be theorised as "embedded ecology",

M. Baindur, *Nature in Indian Philosophy and Cultural Traditions*,
Sophia Studies in Cross-cultural Philosophy of Traditions and Cultures 12,
DOI 10.1007/978-81-322-2358-0_7

where direct ecological themes seem to be absent in the culture, but are found embedded in cultural forms. In a similar essay (Nagarajan 2000b), she adds that such a theory emerges through an exploration of subtle and complex relationships between the "cultural and natural worlds" (p. 454).

Many natural features that are geographically located—rivers, forests, mountains, beaches, stony outcrops, and tree groves—are sacred places in the Indian subcontinent. The relation of people to landscape is articulated in terms of functions, resources, experiences, narratives, and seasonal practices. According to Tuan (1974, p. 146), "sacred places are locations of hierophany". He also points out that places acquire sacred character whenever a divine manifestation or a significant event is associated with it. Many such sacred places in India are described by myths called *purāṇas*. Most of these mythologies are based on the Vedic cosmology. A complex process of place making—by Vedic and Purāṇic primary narratives and localised oral secondary narratives—is usually connected to such landscapes. An examination the cosmology and cosmogony prevalent in many cultures shows how nature in India is perceived from a deeply humanised world view given by creation myths of some sort or the other. For instance, explanations to account for the sacredness of a landscape are often found in the primary narrative called *sthala purāṇa*. Secondary narratives called *sthala māhāṭmya* to recount the human experience of the sacred are also prevalent at most sacred sites.

Inside the world view of such a culture, one can see there is the advantage of not having to work through a relationship with a passive inanimate other-than-human nature, in order to extend an ethical stance towards it. Already present in this culture, is what Abram (2004, p. 83) refers to as "... a common ground, common medium through which a mutual exchange can unfold". This common ground in Indian thought is a pre-existing relationship between human beings and nature as sacred. However, is this idea such an easy "words to world" fit?

In this chapter, I also foreground some of the implications of what it means for nature to be sacred within the Hindu cosmology and culture. Would our cultural reverence of sacred nature result in any ethical framework for actions that protect the planet Earth we live on? The reverence accorded to natural objects or landscapes seems attractive enough to use at least as a metaphor to encourage pro-environmental behaviour in the Indian subcontinent.[1]

7.2 World Views Behind the Sacred

In India, almost all of nature is still perceived from a deeply humanised perspective, given by meaning making that encompasses not just earthly, but larger cosmic realities. Cosmologies, as a meaningful description and understanding of the world,

---

[1]Parts of this chapter were published as a paper in a journal: Baindur (2009).

are very influential and even necessary for a community to understand and interpret their relationship with the universe. A cosmology can be said to be "a theory or conception of the nature of the universe and its workings, and of the place of human beings and other creatures within that order" (Bowie 2006, p. 108). Cosmology, which is often seen as irrelevant or outmoded for modern times, has to be recognised for its fundamental importance within human societies, claims Mathews (1994, pp. 11–12). She writes about the adaptive function of a cosmology:

> Primarily, perhaps, one of orientation—a cosmology serves to orient a community to its world, in the sense that it defines, for the community in question, the place of humankind in the cosmic scheme of things. Such cosmic orientation tells the members of the community, in the broadest possible terms, who they are and where they stand in relation to the rest of creation. Some conception of a cosmic scheme of things is active too in the prescription of a system of norms, or at least in contributing to the normative tone of the community (p. 4).

Cosmologies include within them stories and narratives that explain the world, its origin and the origin of human beings. These myths and stories are collectively referred to as cosmogony. These traditional forms of knowledge help communities understand their complex relationship with the environment and the world they live in (see Bowie 2006, p. 108).[2] Bowie (2006) remarks that Matthews's insights on the importance of cosmology on cultural and social practices cannot be dismissed lightly. She points out that, according to philosophers such as Freya Mathews and others anthropologists such as Roy Rappaport, a cosmology can have both functional and dysfunctional affects. People speak of, act on, and interpret realities that they encounter not as they are, but as they should occur to them or as they are meant to occur within these discourses and narratives. Rappaport (1979, p. 97) too describes the close relationship between nature and cosmology:

> Nature is seen by humans through a screen of beliefs, knowledge purposes, and it is in terms of their images of nature, rather than of the actual structure of nature, that they act. Yet it is upon nature itself that they do act, and it is nature itself that acts on them, nurturing or destroying them (p. 97).

It is important to mention here that the world views of the people may not be rational or scientific enough to support the perspectives of modern conservation, but such beliefs or presuppositions may as well serve to encourage actions that further environmental causes. In fact, we find that the religious and cultural associations created by such narratives often have more popular impact than the tangible values of resource or the discourses of environmental conservation. Alley (2000) writes "Symbolic representations of space in Hindu sacred texts and ancient concepts associated with them call for an approach to ecological understanding that moves beyond the secular notions of the 'environment'" (p. 298).

Two of the popular themes that invoke the idea of sacred nature in Indian thought are that of sacred geography and the equally significant discourse of the Goddess,

---

[2]While cosmologies are theories that relate to the origins of a universe according to a world view, cosmogonies include stories and myths too.

Mother Earth, or *bhū devī* that has been discussed in Chap. 6. Both these themes underlie popular imaginations, rituals, discourses, and textual literature around nature in India. Kinsley (1995, pp. 58–60) foregrounds the idea of land as a holy place as being one of the predominant ecological themes of the subcontinent. He suggests that the sacred geography of land manifests itself in narratives of the Earth goddess, sacred geographical features, as well as the theme of sacred Bharat-*ma*, mother India. The first of these two beliefs is certainly ecologically relevant and is analysed in this chapter. The prevalence of conceptualisation of land as sacred and the Earth as both mother and sacred goddess suggests that the Indian geographical or terrestrial experiences of nature are interpreted through cosmogonies that are metaphorical and deeply linked to sacred myths. One reason why it is important to look at these two themes in particular is because these ancient narratives and cosmogonies form a living tradition practiced in the everyday lives of people.

What is the source of the evolution of cosmogonies behind these kinds of themes and are such cosmogonies are relevant to creating a background for some kind of ecological ethics? The increased interest in understanding ecological beliefs in non-Western cultures is spurred on by a renewed interest in the concepts that are available in traditional Indian thought. Reverence to nature is often cited as a possible virtue we could re-learn from our pre-modern past. Many writers make the claim that a large part of nature is considered sacred and worshipped in India; therefore, it could have positive implications for an ecological ethics of nature. The re-reading of these textual sources in the current literature has the purpose of finding possible concepts (often referred to as conceptual resources) that could be used to generate some form of an indigenous environmental ethics. Moreover, it is clear that attribution of value and sacredness to these objects is dependent on many factors, including folk narratives, and is not limited only to the larger Hindu traditions.[3] Nagarajan (2000b, p. 281) explains how this arrangement of the cosmos occurs: "Depending on caste, class, religion, community and bioregions, different people arrange natural substances according to a diverse range of values".

One of the reasons we value something is because of the way we are related to it. If we can find such a way of relating to the nature as Earth or relating to sacred landscape as a vulnerable form, it would then justify the view that conceptual resources from Hindu thought can be used to raise ecological consciousness (within our own country) in the current scenario, and also yield alternate frame works of values and ethics that are not mere imitations of Western traditions of moral and ethical thought.

[3] It is, however, an onerous historical task to find out whether these folk traditions influenced the major texts or vice versa. It is clear that there have been exchanges at various periods of Indian history.

7.3 Sacred Places and Natural Features

7.3.1 The Idea of the Sacred

A brief note on some perspectives of the idea of sacred in Indian thought is discussed here in detail to provide a background to the idea of "sacred nature" in India. Among objects that are related to the practice of sacred nature worship or reverence, we must distinguish between natural objects, general geographical features, and particular *named* landscapes. Certain objects, plants, animals, and features are regarded as sacred categories—such as the *tulsi* plant (*Ocymum sanctum*) or the holy fig tree (*Ficus religiosa*) or eastward flowing river channels. The sacredness of these objects is given by the fact that they are sacred types and tokens; they are considered as some sort of sacred universals. They are not bound to a particularised location. All occurrences of such objects are sacred. On the other hand, there are objects in which the divine is invoked for some time and the object is sacred for that period of time. Clay idols of Ganeśa used in the annual worship in homes are sacred after they have been installed.

To distinguish the idea of sacred objects, one can examine the way the divine is invoked in such an object. Keith (1925) points out that there is difficultly in distinguishing between the divinity accorded to the sacred by being imbued with the holy or sacred essence and the object itself being divine, a reverence paid to the sacred object as a sign (see Keith 1925, pp. 1–66). This difference is not confined to the natural but also to human-made objects. For example, the various ceremonial objects in Vedic rituals such as the wheel refer to the sun and gold or sometimes represent the god Agni or other deities (pp. 66–67). Yet again, the stones with spiral markings—the *sāligrāmā*—are considered divine forms of Viśṇu. Sometimes, the objects used for a sacred ritual may not be used or treated like ordinary objects. Flowers used for ritual worship are ritually divine and may not be disposed with normal garbage even after they are dry. Strings of used garlands hanging from trees in streets in urban areas in India are an illustration of this point.

An interesting idea has been proposed by Apffel-Marglin and Paranjuli (2000) about the notion of sacred (see pp. 291–316). They argue that the sacredness cannot be equated with a non-utilitarian attitude. In other words, the attitude of reverence within Indian thought does not presuppose an attitude of "non-use" by human beings. This is a very relevant insight for the purpose of understanding the practice of reverence. They posit a paradigm of rest/fallow and active/productive phase of the use of the Earth and land (p. 305). Revered objects within Hindu cosmology cannot be totally kept apart from the people. Depending on social practices, ritual beliefs, or other kinds of rules, there are restricted areas of the sacred that one may access, but in general, access in terms of its sacred function is not denied.

7.3.2 Sacred Geography

A traditional belief system of reverence for landscape features, referred to as sacred geography in popular literature, is common to most cultures in the Indian subcontinent. Some writers also suggest that forms of mythic–ritual sacralisation or reverence of geographical features could translate into ecologically supportive behaviour by the people. Ruether (2005) points out the prevalence and the intensity of these beliefs and their connection to ecological ethics in the Indian subcontinent. She writes "Perhaps nowhere is there such an extensive sense of sacrality of place. Its [India's] forests and rivers are seen as holy, even as embodied gods and goddesses. India as a whole is venerated as a sacred land" (p. 48).

With reference to the beliefs about sacred features, we find that each of these sacred places has a rich narrative tradition, either oral or written that describes in detail the story of sacred origin of these places. Serving as both markers of events and as metaphorical teachings for everyday behaviours, the stories called "Purāṇas" or history form an important component of the philosophic presuppositions about the concept of nature. Tuan (1974, p. 146) suggests that the landscapes become sacred after a signal event demonstrated by the occurrence of some signs: "In every instance the spot was sanctified by some outside power, whether it be a semi-divine person, a dazzling hierophany, or cosmic forces that undergrid astrology and geomancy" (p. 146). In the Indian traditions, *purāṇas* relate these events and also describe the context for the appearance of these sacred signifiers. Such stories are not limited to the larger textual tradition of Hinduism as we know it today, but they are also told from a very local context and through oral or ritual folk traditions. These local narratives such as the stories of the land are deeply embedded in the geography of a place.[4] Chapple (2000), for instance, writes of such narratives: "It must be noted, however, that many pilgrimage places within India, from the Himalayas in the north to Kanyakumari at the very southern tip of the subcontinent, form a patchwork of sacralised spaces that could be newly interpreted through the prism of environmentalism" (p. 33).

To make the idea of sacred narratives clear, I retell one of the stories from the south of India, about a sacred landscape, a hill called "*maruda malai*" or medicine hill here. Rare medicinal herbs are found on this mountain near Coimbatore in South India, which is worshipped as a sacred geographical feature.[5] The story of how the scared hill came to be located in that region is connected to an incident that is narrated in the epic poem *Rāmāyaṇa*, the story of the divine hero Rāma.

During the final battle described in this epic that is said to have happened in Lanka (Sri Lanka), *Lakṣmaṇa*, the ideal brother of the divine hero, Rāma, was lying on the ground senseless bound by the magical bonds of the *sarpāstra* (a mythical snake weapon). The only cure was a divine herb *sanjīvinī,* found beyond the Himalayas.

---

[4]In time, however, such local stories get absorbed into the larger mythologies of Hinduism or adapt to their own versions depending on the socio-political views of the people belonging to a place.

[5]The name is in Tamil, the local language. For details of this story, see Das (1964, p. 6).

Hanumāna, the mighty *vānarā* (monkey warrior), flew north to the Himalayas to find the herb. Arriving at the mythical mountain (also called *sanjīvinī*), he found that all herbs on the hillside were alike and he could not identify the right herb. So, he picked up the whole mountain in his mighty hands and flew down south towards Lanka. On the way, a piece of this hill fell down in the south of India; the sacred hill called "*maruda malai*" is said to be that very piece. According to the story, the expert physician in Lanka identified the herb and revived the ailing warrior.

This story may or may not be available in the Sanskrit version of the *Rāmāyaṇa.* But "natural sacred places" consisting of geographical features are revered and are almost always associated with oral narratives about the location called *sthala purāṇas,* and this story recounts one such narrative.[6] The shared meanings and communicated oral histories of natural-scapes draw back to the deep connection of nature with the concept of earth, or land (*bhūmi* in Sanskrit). The idea of these oral histories therefore is not only embodied in architectures of the human being, but also includes natural elements and natural objects, specially water, rocks, and trees which form features such as rivers, lakes, mountains, and forests. These locations are not *universal* places or generic features such as sacred groves or river confluences, but are *particulars* (specific to their cartographic positions). The particular and unique nature of each feature is given by mythical imaginations of journeys, of events, and of creation. Many ritual practices that may be religious or cultural are performed in these areas to reinforce the narratives again and again.[7] The myths answer the question as to why the place is sacred and often give a name of the place that is based on one of the themes of the myth. In such narratives, often divine and superhuman events are described and are claimed to have happened in a particular place. These are recollected by a set of oral stories narrated by the local inhabitants, or mythical histories sometimes visually represented during festivals through folk performances. Eliade (1959, p. 95) refers to the power of such myths to create what he calls "an apodictic truth". He asserts that the myths create a reality by revealing a sacred history: "It is the sacred that is pre-eminently the real". People who encounter the tangible elements of such natural landscapes do not see them as sterile nature or mundane phenomenon, but perceive them as sacred locations and experience the sanctity of contact with the place. What I derive from this reading of Eliade (1959) is the converse idea that perhaps a natural spot that is not construed with a myth or sacred history is mundane. This has some implications for environmental ethics that will be discussed later.

---

[6]In Sanskrit, *sthalā* means place, *purāṇa* means history or ancient stories, so the word would mean ancient story of a place. These have been documented and published into written books only in recent times. Sacred groves also belong to this landscape category with their own stories.

[7]Annual festivals, rural fairs with folk narrative, and drama traditions often portray the narrative at the sacred place.

7.4 About the Idea of Place

To examine the idea of the sacred natural landscapes in India, the concept of "place" as described by Edward Casey (1993) is useful. While the notion of "space" represents a three-dimensional, measurable extension of elements grouped together, at a more experiential level, place itself would include the "character" of the space, which one can loosely term as the social and culturally defined space. According to Casey (1993), the power of a place is not merely determined by its location on a map, but includes the relationships of the elements within it.[8] He writes:

> The power of a place such as a mere room possesses not only where I am in the limited sense of cartographic location but how I am together with others (i.e., how I comingle and communicate with them) and even who we shall become together (p. 23).

This idea of place certainly allows for the rich connection between a habitat and its occupant more descriptively and completely. Due to the importance given to notion the human interpretation of and interaction with the natural areas within the Indian subcontinent, one has to examine the nature of "implacement". Casey (1993, p. 23) reiterates that "implacement is as social as it is personal". We find that on the one hand, the notion of sacred is personal and cultural in Indian thought: people often visit the sacred places for their own spiritual or personal benefit. On the other hand, the social and cultural beliefs about the sacredness which guide these visits are oral narratives that are socially shared. As a phenomenon of experience, the feature itself is natural, an area on the geographical landscape—the contact with the sacred is very personal, but the story about it is cultural and intersubjective.

It is to be noted that, in Sanskrit, the idea of place has two equivalents. The conceptualisation "place" in Sanskrit is thus deeper and more specific in its interpretation and meanings. The first term "*sthala*" is often used to indicate place as an area on the ground. The word "*sthāna*" which is more like the word "spot/place" refers to a designated location and is a term that also performs an indexical function.

The narratives and stories that surround the natural can only be understood with a concept that connects the idea of the sacred and the natural, a concept that intimately connects a human being with her environment. Unlike the word "place" that can be used to designate place order and other locative references, the use of the Sanskrit term "*sthala"* is free of a mere indexical function. *Sthala* in Sanskrit refers to a section of the Earth that is distinctly marked out from the rest of the landscape as different. *Sthala* is derived from "*sthā*", meaning "section, chapter, or marked part" "and" "*talā"* which means "surface". The correct translation would be "land-place". The surface of the Earth is intrinsic to this word. That eliminates for us the possibility of having to include places otherwise not found on Earth such as location in the sky or clouds. The places in heavens (*swarga*) are never called *sthala* in the mythology. They are sometimes called *sthāna*, such as "Indra-*sthāna*". Distinguished from the other word *sthāna*, referring to the ordained **location**,

[8]Parts of this section are from Baindur (2010a).

or where things stand, or can be spotted, *sthala* is a terrestrial-linked term and a shared cultural kind. For example, in the *Tīrtha Yātra Parva* of the *Mahābhārata*, a sage describes the holy places to be visited and the merits gained by the pilgrims:

> "*tato gaccheta dharmajña viṣṇor sthānam anuttamam*"[9]
> Then, (one should) go to the most exalted place of *Viśṇu* (where he is established), O knower of righteousness... [10]

We find that in common usage, the sacred landscapes that are sacred places are referred to as *tīrtha sthala*. The designated location of a city/temple or the location in some area is designated by a different word, *sthāna*. I argue that the word "place" therefore is closer to the word *sthala*, especially when it refers to a landscape or more accurately a land-place.[11] Besides, we find that *sthala* is invariably linked to the material boundaries of a landscape feature such as a rock, a river, or a forest. The sacred boundaries of a place are as amorphous as the landscape they designate. Even in the case of *place*, Casey (1993) points out the difficulty in the distinction between *place* and *landscape*: "A landscape seems to exceed the usual parameters of place by continuing without apparent end; nothing contains it, while it contains everything, including discreet places, in its environing embrace" (p. 25). The same holds true somewhat for the category *sthala*. But discreetness of *sthala* for human recognition, however, is given by the material content of the feature, such as the water of the Gaṅgā river, the red soil of a particular region, or the extent of a rock surface. Though most sacred places involve some form of a water body, the category *sthala* includes the area in general, along with the banks and the land surface. Where discreteness of a boundary is lacking, the sacred place is marked by human architecture, such as steps on the river bank or a shrine marker which indicates the horizon or boundary of the sacred place or simply serves as a pointer. Tanaka specifies that sacred sites such as these "comprise natural and human-made assemblages of sacred symbols and landscape markers invested with special meaning" (Tanaka 1988, pp. 21–40). It is clear that it is the natural feature itself that is sacred and not just the shrine. As mentioned earlier, the word *sthala* seems to be closer to the idea of a natural feature because it includes the idea of "land surface" within its interpretation and so it is the carrier of the created sacred reality. As we have seen in the earlier story, the "hill" or "*malai*" is the natural meaning and so is the word "*maruda*" which refers to the profusion of medicinal herbs on the hill. The origin of the hill and its divinity are explained by a narrative. When some of these narratives are analysed, we find that there is a certain world view of mythical history associated with these geographical places—a narrative of ***being*** sacred by creation, rather than being ***made*** sacred.

In a conceptual analysis of sacred geography, while examining the ideas and practices that surround the concept of the sacredness of land-place, two components

---

[9]Emphasis in bold is mine.

[10]*Verse 10, Chap. 83, Vanaparva, Mahābhārata*. Trans by author.

[11]The word "landscape" devoid of its historical antecedents in the west would be ideal as the translation as it includes within it the word "land".

of experiences of the people can be identified: firstly, the mythic imagination which relates to the sacred origin of the land-place and, secondly, the ritual practices that are prescribed in such places. The oral narratives that record the sacred origins of the place or *tīrtha* are often called *sthala purāṇa* (story of the land-place). What is unique to these stories is that along with the meaning ascribed to natural objects or elements of the environment, each place is connected with a story that is rich in metaphor and includes the location and its natural elements within its narrative, along with people and divine beings. Eck (1990) mentions the prevalence of these stories: "The stories of India's *tīrthas* are told in the popular praise literature, the *māhātmya*s, sometimes called *sthala purāṇas*, 'the ancient stories of the place' (p. 35)".

Sacred *sthala* narratives have two components: one is a description of the descent of the divine upon the terrestrial or "earthed divine" and the second part describes human experience of this divine on Earth or "deified land". The secondary narratives, the *sthala māhātmya*s, describe the positive interactions of people who have benefited from the presence of the sacred or have been punished for disrespecting the location. These secondary narratives are often referred to as *sthānamāhātmya*s too, especially if there is a temple or a shrine. For example, in the famous shrine of Somanatha, the oral *sthala purāṇa* recollects the manifestation of a self-formed *liṇga* or a representation of Śiva at a river confluence.[12] The *purāṇa* tells the story of the moon god, *Soma,* who by being partial to one of his wives of the other twenty seven sisters he married annoyed his father-in-law. Cursed to be consumptive, the moon was unable to perform his duties. To restore his brightness, he was asked to bathe at the confluence of Sarasvatī. The *Skandapurāṇa* states that Sarasvatī originates from the water pot of Brahmā in the heavens and flows from *plakṣa* on the Himalayas. The myth speaks of how by bathing at the confluence of the rivers; he regained his splendour and had a vision of *Śiva as* a self-formed effulgent *jyothir-liṇga* (a *liṇga made of light).* The term "*prabhāsa tīrtha*" is named after the regained effulgence of the moon. Both non-earthly entities, the *jyothir-liṇga* and the descent of the celestial moon itself into the waters, further sanctified the holy place. The *sthalamāhātmya* recollects the association of Kṛṣṇa with this place. Also popular is the story of King Mūlaraja of the Chaulukya dynasty who built great shrine at this place after a dream about the moon god. Thus, the two forms of narrative coexist, informing the pilgrim that her experience of the holy place is sacred and otherworldly. To quote the words of Flood (1993): "Mythical worlds are mapped to specific geographies of a holy place; the physical world is imbued with mythological or religious meaning" (pp. 1–5).

Some spaces or areas on the ground are "originally sacred", while others are sanctified by rituals of human beings or actions of the divine beings. For example, before building any structure, the land is consecrated and worshiped with the ritual of *bhūmi puja*, or land worship. There are rituals where land areas are temporarily

[12]For a detailed mythical history and the story of this shrine which is condensed here, see "The Setting", Chap. 2, in Thapar (2004, pp. 18–37).

sanctified for a *yajña* (Vedic ritual) or a *pūja* (worship). On the other hand, the idea of naturally sacred locations is interpreted through narratives that claim sacredness for the land-place by some sort of ***non-terrestrial*** "origin". Land-place features, however, are connected to very specific, particular examples of sacred events that have occurred in an ancient time and space. With respect to natural elements, it seems that both kinds of sacred narratives exist—divine origin and divine contact. There are areas and sacred places that are originally sacred and some ordinary places which are made sacred by connection with the divine. It is to be noted that the narratives of places sanctified by contact with the divine are not unique to Indian thought alone and also that the idea of divine contact is not restricted to natural landscapes, but includes human-made objects or even relics.

While the rituals of purity or actions of the divine gods create sacred spaces, geographically sacred regions are ***implanted*** onto the Earth. These regions seem to have sacredness as an essential component. The sacredness imbued in the landscape features—rocks, mountains, or rivers—does not disappear after the human or divine interaction is complete. The sacred spaces that are created by ritual acts may later turn mundane, while sacred places remain sacred, regardless of time and changes. Within the belief system of *purāṇas*s, the defilement of a sacred geographical feature is not possible, making the environmental efforts around these natural features a difficult task. We therefore need to understand what makes these sacred regions incapable of being polluted within certain belief systems of Indian people. Here, I propose a radical idea that perhaps it is the narratives about the origin of these places that makes them non-degradable. I suggest that sacred geography is not geography of "terrestrial", but of implaced other-worldly materials—rivers, mountains, or forests.

7.5 Sacred Imaginations: Myths About Sacred Places

Most myths about the sacred places are a narrative about the transplantation or a sudden appearance of that sacred feature on the Earth. This narratives or stories are like mini, creation myths and discuss the divine origins of the sacred land-place. These narratives form a part of the tradition of stories called *purāṇa*s. The descriptions of nature-scapes and the relationship between the human and nature in the Vedas and the *purāṇa*s, some of India's earliest philosophical and religious literature, are to be understood within a broader framework of some fundamental conceptions of people who created these narratives and their larger cosmic views.

The first of these preconceptions that I discuss here is that of the idea of nature itself. The popular meaning of nature in the current times is that which is "non-human". It is also clear that the idea of a "non-human nature" is largely absent in these stories and narratives. However, the perception of nature is anthropocentric; much of the manifested world is explained and understood through the experiences of the human being. Human beings are not placed above all other natural elements in the world, but they are situated in the cosmic system, interrelated to both beings and geographies. Bilimoria (1997) stresses on the cosmic application of moral values across all beings:

> The normative values were not restricted for human well-being alone, rather they were universalized for all sentient beings and inanimate sectors as well as spirit-spheres, i.e. gods and the faithfully departed; the biosphere, i.e. animals and plants; and the broader biotic universe, i.e. inanimate realms comprising the elements, stones, rocks, earth-soil, mountains, waters, sky, the sun, planets, stars, and the galaxies to the edges of the universe (this and other possible ones) (p. 2).

The classificatory scheme in case of Hindu cosmology is based on a cosmic system of the place-worlds that are called *loka*s, *and* each *loka* is an inhabited world each with its own description, having within it unique features, denizens, places, myths, and also creation myths within a cosmology. The Earth itself, as the **terrestrial** surface which humans and other Earth beings occupy, is not seen as a single isolated place, but it exists as part of a hierarchical cosmic system of different *loka*s or worlds.

Thus, we find the worlds, though placed in relationship with each other, are still bound by an order of divinity and importance. The heavenly world, being relatively immortal and replete with pleasures and privileges unavailable upon the Earth and other "lower" worlds, was accorded a higher status value. To be a denizen of these sacred worlds required actions and austerities that naturally made a being of a world morally higher in status. It is significant to note that at the cosmic level, beings were not ordered according to families or species; ordering was first based on origin and the world they inhabited. Thus, a divine serpent is higher in the hierarchy than the human being on Earth. The achievements of pious acts or austerities also mattered in this moral order, sometimes being considered more important than age or the place-world of origin. Many stories recount how sages from Earth, for instance, were revered by even the gods. The understanding of this world view is significant to the understanding of the creation myths of sacred landscape features on the Earth.

7.6 Crossing Over: Places and Human Beings

As in the case of the *maruda malai* or the sacred medicine hill, it is clear that the sacredness of the landscape is connected to the origin of the event that caused the hill to *be* or *occur*. Accordingly, in story of the medicine hill, it actually dropped from the sky onto the Earth. The unique creation of land-places in these mythological narratives strongly support a hypothesis that divine origination alone imparts eternal sacredness to a land-place. I suggest here that the purpose of the narrative is to locate the place culturally as non-terrestrial (non-earthly) and give it a higher value than that of the surrounding areas.

The reason such places are sacred and divine is because the feature is not earthy, but has been introduced by an event from another *loka* onto the *bhūloka*. The creation of the land-place feature more often than not signifies a geographical descent of some "other-worldly feature". The descent (not fall) from the higher worlds is easier than the ascent. The descent of anything from the divine plane forms an important part in the creation of the sacred in nature. The descent of the divine or *avatār* concept that is very much a part of the Vedic and Purānic tradition finds its counterpart in geographical descents of rivers and other natural features onto the terrestrial. By being descended, these transfers of rivers, mountains, lakes, and rocks from the heavens help the terrestrial beings, such as the human, ascend.

The primary goal is the gaining of positive *karma* that allows one to access higher births or planes such as heavens, as well as the opportunity to attain *mokṣa*, the cessation of suffering. The transit between worlds is possible for the beings that have eligibility or have gained enough merits (good *karma*). Eck (1981) sums up the idea that *tīrtha*s are like ladders to higher worlds: "In sum, it is clear that the *tīrtha* is not only a riverside bathing and watering place, but a place where one launches out on the journey between heaven and Earth. It is a threshold of time, or space, or ritual" (p. 328).

The original sacred thus comes from the heavens—the *devaloka*. The rivers of India form one of the most striking examples of this origination as sacred narrative. The *Ṛg*-Vedic myth, in which *Indra* slays the serpent *Vṛtra*, who had coiled around the heavens and locked the waters inside, and thus frees the heavenly waters to fall to the Earth, is recounted in this verse: "As your ally in this friendship, *Soma*, *Indra* made the waters flow. He slew the serpent and sent forth the Seven Rivers. He opened, as it were, the holes that were blocked" (trans. Griffiths 1973).

Though these narratives of direct descent are far and few, it seems that there are many more features that somehow are accounted for by oral histories that may not occur in the literal rendering of the Purānic or Vedic texts. Historically, it seems likely that these located sacred land-places were adapted from an earlier primitive tradition of spirits abiding in nature. Eck (1981) writes:

> … the many specific *tīrtha*s of India's vast sacred geography are also well grounded in yet another tradition: the non-Vedic tradition of indigenous India which, despite its many areas of obscurity, was most clearly a tradition of life-force deities associated with particular places. It was a locative tradition in which *genii loci* under a variety of names—*yakśas, nāgas, ganas, mātrikas*—were associated with groves and pools, hillocks and villages, wielding power for good or ill within their areas of jurisdiction (p. 324).

She suggests that the traditions of pilgrimage by foot or sacred journeys are traditions based on sacred place. She adds that these myths are not static and keep changing, yet the places draw pilgrims who come to presence the divinity and seek blessings from the resident deity (p. 324).

Eck (1981) refers to this as borrowing and assimilation of the pre-Vedic tradition into the *purānic* lore. It also seems likely that many places create the narratives that give them legitimacy through the association with popular Hindu texts and gods. Often, in its *māhatmya*, a local *tīrtha* will subscribe to the larger all-India tradition by linking its sanctity to the great events of the major epics and *purāṇa*s. She suggests that this might be seen as the geographical equivalent of Sanskritisation (p. 336).

7.7 Sacred Interactions: Human Aspirations

The story of the descent of the Gaṅgā is much eulogised, having many versions and subplots within the main story. In all the versions however, the narrative implies that the actual river, materially, is not of the Earth but of the heavens and is of godly content and essence. The presupposition that makes this transfer of material possible from one *loka* is that the substances—gross or subtle—are all the same and are made of the five elements. So, a river from heaven is as real as one on Earth. But, its reality is a sacred reality, not the reality of the Earth. The way this river differs from an ordinary earthly stream is by having the quality of sanctifying human beings and the earthly plane, and her origins from *devaloka*.

In the secondary narratives of the heavenly river flowing upon the Earth are recounted the various miracles wrought on the human beings who take a dip in her waters. The claim is that the experience of the ritual dip (ritual bath called *snān*) is a terrestrial experience of a dip in a heavenly river that has been transplanted to the earthly plane. The interaction between the land-place and the human pilgrim in his embodied form can be conceptually understood by using the concept of place as theorised by Casey (1993).

The human subject gives identity to the undifferentiated geographies of a landscape or natural regions by her interaction with the phenomenon and ordering them into fragments of private and collective memory. The experience of the human in the sacred natural land-place is different from the experience of a human being in a sacred place like a temple. This seems to be an example of what Casey (1993, p. 31) calls a "placescape". He refers to a placescape as something that is generated by a collusion of the body and the landscape. This identification of specific locations into placescapes occurs each time the subject comes across unfamiliar territory—natural or settled (p. 31). By this definition, sacred land-places are placescapes because they are created by a collusion of the Earth beings and land-places that have originated from the divine worlds. Though located on Earth and near enough to the familiar human habitats, the land-places by the nature of their origin are alien, unfamiliar. The narratives emphasise contact of the divine material with the body in the sacred place, rather than give priority to the experience of the presence or "*darśan*" through a symbol or vision. The importance of bodily contact with the divine reality is both phenomenological and ontological. The acculturation of these landscape features, according to Casey (1993), is "a social or communal act". Place as the sacred landscape thus is no longer just a "natural" category, it includes within itself a historical component. On how these places both cultural and social become shared realities, he writes "The culture that characterizes and shapes a given place is a shared culture, not merely superimposed on the place but part of its very facticity" (p. 31).

The experience of the human being who has bodily contact with a part of the divine world is very much linked to the idea of *karma*:

The dust (*dhūli*) from a sacred place has a special significance for a vaiṣṇava... While visiting the *tīrthas*, the pilgrims rub the dust of the holy place on their forehead and body as a mark of humble devotion (Chowdhury 2000, p. 74).

The *sthala māhāṭmya* story of the Pāpanasaṃ Waterfalls (in Tamil, the word *pāpanasaṃ* means destroyer of sins), further illustrates this point. A brother and sister separated at birth by calamity fell in love with each other by mistake. Soon, they both realised that they had sinned and wished to make amends by visiting all holy rivers and waters. Learned people advised them to wear black garments and bathe in all the holy waters, until the clothes turned white. No holy place gave them any relief, until finally they bathed at the waterfall called Pāpanasaṃ. On bathing in the falls, their clothes turned white and they achieved salvation. The fish that live in the lake are golden-hued and are never killed or eaten (Das 1964, pp. 44–45).

Whether it is the contact of the mud, water, land, or herbs, with the body of the devotee, the natural *tīrtha* is much favoured over the built structures. Perhaps this is the reason why many temples claim that the image of god was "found" rather than made. For example, the famous statue of Lord Bālaji in Tirupati is said to have been dug out from the earth by a devout king. This suggests the image was not of human origin but "other worldly"—a direct descent of the lord from his divine world in a corporeal image form. Naturally occurring Śiva stones or the Śiva *linga*s are also said to spring from Śiva *loka*. Referred to as *svayambhū* (self-born), they attract worship in the most unobvious places even today such as an urban horticultural garden or in an ice-sculpted form, or in the holy mountain shrine of Amaranth (where a Śiva *linga* of ice is formed annually), reached after an arduous trek. This tale also demonstrates Eck's (1981) explanation of the concept of *tīrtha*, or crossing over. Every sacred location forms a ladder, where the human can crossover to the state of salvation or to a state of heavenly experience of purity.

Eck (1981) remarks on the living tradition of these narratives: "The whole of India's sacred geography, with its many *tīrthas*—those inherent in its natural landscape and those sanctified by the deeds of gods and the footsteps of heroes—is a living geography" (p. 336).

Place as a natural landscape includes time as an integral component of happening, not marked by physical parameters but by the experience of a subject. These form the basis of both shared and unshared narratives. Most rituals and stories associated with the place can be dismissed as mythical, but they are deeply metaphorical and give insights into the place-experiences of these traditions. In Casey's words: "We might even say that culture is the third dimension of places, affording them a deep historicity, a *longue durée,* which they would lack if they were entirely natural in constitution" (1993, p. 32).

7.8 Issues Around the Sacred Places: Being Immutable

Land-place features are sacred, yet the reverence seems to be merely ritualistic without regard for the physical degradation of the natural. A dichotomy between the sacred and the mundane that Kinsley (2000, pp. 225–246) refers to exists as two different spheres of belief. While the sacred landscape affects and impacts the human beings, the lower valued human being has no impact on the sacred in return (in comparison with the more sacred and exalted status of the divine-worldly land-place). On the other hand, the mundane activities of the human are both impacted and in turn affect by the natural—as in case of the pollution of the Gaṅgā or the destruction of a sacred grove. What are the understandings of the sacred and the mundane with respect to natural landscapes? There exist some conflicting notions of sacred that I intend to discuss here.

From the conclusion about the origin myths of the sacred *tīrtha*s, one can reason that the sacred is immutable and the attitude towards the sacred is one of ritualistic reverence, not environmental restraint. Alley (2000, p. 322) writes about the two conflicting notions of pollution that exist in the Gaṅgā. It is seen that the ecological idea of pollution relates to chemical and other scientific parameters, while the priests equate the impurity to break down of morals and social values. The idea of the sacred land-place is located in the sphere of the sacred reality, not the mundane world of water and dirt. The original sacred, therefore, is considered immutable and cannot be subject to degeneration. As I mentioned earlier, it is also true that whichever land-place is not construed with a myth or sacred history, conversely, is mundane. This is an important issue related to sacred natural places that are local in nature. People from different areas who are unconnected to a sense of the sacred place lacking the experience of the shared narrative would not believe in the local sacred geography. In an essay comparing the pilgrimage of the Hindu with the aboriginal walkabout, Kingsley (2000, p. 228) points out how the sacred myths are like the dreamtime tales of the aboriginals and the landscape can be imagined as a text containing a detailed narrative of the land, in which these people are embedded. He suggests that these implicit structures are not comprehensible to a person who is outside the cultural context, and such a structure plays more than a mere geographical role (p. 229). This idea suggests that it should be the local carriers of the sacred myth, who should be the enforcers of any plausible ethics of place. Since the sacred is already embedded in their practices, including the ecologically relevant ethics would be easier.

The argument by environmental philosophers is that we do not have a theory of ecological ethics in Indian thought, but only have a kind of a normative framework that can be called at the most descriptive. Merchant (1980), for instance, emphasises that the normative import of the descriptive statements of nature is also important. She argues that understanding changes in the description of nature could lead us to understand certain ways in which cultural values have changed. She also implies that such descriptions lead to exposure of hidden norms within a culture: "Descriptive statements about the world can presuppose the normative; they are then ethic-laden … the norms may be tacit assumptions hidden within the descriptions in such a way as to act as invisible restraint or moral ought-nots" (p. 4).

Gottileb (2004, p. 8) also emphasises the significance of texts that teach systems of beliefs and create identities for the human being that seem to go beyond merely social or physical identities. Though most of these narratives seem to be about other-worldly concerns, they play a very crucial role in people's orientation to their everyday world especially in the form of familiar habits and rituals: "At the same time religions provide norms for the conduct for the familiar interpersonal settings of family, community and world. Religious moral teachings presuppose a spiritual foundation and are meant to root our everyday behaviour in a spiritual truth about who we really are" (p. 8).

7.9 Conclusions: The Problems and Possibilities of a Sacred Nature Discourse

The idea of a scared geography can contribute positively to environmental ethics. Along with the discourse of the sacred imagination, the secondary narratives include normative rules that are to be followed in sacred places. Like the restraint on fishing in the holy falls mentioned earlier, many types of rules also surround the conduct of pilgrims to a sacred place. Jacobsen (1993) calls these two discourses as the magic and the ethical discourse and emphasises the importance of the ethical discourse: "The second group of textual statements aims at having an ethical impact from the point of view of environmental ethics of the place" (Jacobsen 1993, pp. 141–149). Illustrating the importance of the practice of normative ethical restraints in a sacred place, Jacobsen recounts how pilgrims practice forms of *ahimsā* or non-violence in sacred places by not using footwear or consuming meat. The sacred is to be experienced by morally dealing with the mundane even within the mind. Within the sacred, we do have two schools of thought: one which emphasises that the mere ritual can be sufficient for the benefit of the sacred experience and the second which hold that rituals without the support of moral conduct would not benefit a pilgrim. The popular story is told of how all the sins get off and wait for the bather to take a dip in the Gaṅgā and re-join him as he steps out of the divine river. Such narratives included in the secondary narratives seem to actually critique the sacredness of the land-place and emphasise moral conduct as a prerequisite for the experience of the sacred place. Verses in the *Mahābhārata,* (*Vanaparva*), for instance, describe the various moral practices for an individual that would give him the full benefits of encountering the sacred. They include observances such as self-control, being truthful, following austerities, and treating all beings as he would himself (Kane 1973, p. 562).

There are also ritualistic practices that have a moral basis that seem to prevent pollution of sacred places. For instance, the *Śiva Purāṇa* has a list of practices to be followed near holy water bodies and rivers which includes not spitting into the water, not washing clothes in a river directly, but using the water to wash elsewhere.

While I do agree with the comparison, I would like to point out that mere sacred imaginations of the land-place will not directly contribute to the conservation of such places. Instead, what would have an impact would be an emphasis and a re-awakening of the ethical discourse of restraint that runs parallel to the sacred stories of the land. In the words of Jacobsen (1993, p. 138): "Places of pilgrimage are places where people, according to the normative statements, are expected to show restraint towards all living beings. There is therefore traditionally a relationship between environmental ethics and sacred places".

Despite significant environmental campaigns in the media and the incorporation of environmental sciences into education, there has been no significant change in people's behaviour towards the environment. Recent studies have shown that mere awareness or education on issues does not transform human behaviour. The need to raise these supernatural, socio-ethical beliefs and values to the secular understanding is possible through linking these values to sustainable, pro-environmental behaviour of people. If the value–behaviour link is clearly established, it is possible to incorporate value education into environmental awareness programmes and advocacy. It is, for example, possible that somebody is quite aware of the fact that his behaviour is detrimental to natural environment, but as long as he is not convinced that it is important to preserve the environment he might not be willing or motivated to change his behaviour.

I end this chapter with a very brief note on the possible ways in which the idea of sacred geography can be relevant to ecological ethics. It is in the body of human being that both the mundane and the sacred meet. The human being is the agent of moral action both for the ascent into higher worlds and the preservation of the nature in this world. Though the purpose of the sacred is to create a way for the ascent of the human being, and not ecological conservation, it is clear that the emphasis on the restraining or other similar normative practices can serve to create ecologically sensitive pilgrims. The idea is to include the normative values within a place without the ecological value displacing the sacred value or imagination. Wherever possible, ritual practices must be supported by ecologically planned structures. [13]Asking people not to bathe in the holy waters or visit a sacred rock would not be possible. However, asking them to not use plastic papers or eat or spit within the sacred perimeter would be well within the discourse of a sacred place. It is important to therefore take into account the narratives and concepts of sacred land-places and perceive them beyond the mere natural features to create a viable eco-ethics of place. The relationship to place both socially and culturally given by normative narratives would thus be environmentally relevant in today's world. There are parts of our pre-modern tradition that we have to reject and parts of it that we have to incorporate in this reformed world view.

[13]For instance, in Talakaveri, the spring considered the birth place of the Kaveri, the bathing area is kept separate from the actual spring where worship is offered. *Kalyanis*, or special tanks, were constructed on the lake banks in Bengaluru to provide for the immersion of *Gaṇeśa* clay idols during the annual festival which would have otherwise polluted the lakes and tanks.

References

Abram, D. (2004). Reciprocity. In B. V. Foltz & R. Frodeman (Eds.), *Rethinking nature* (pp. 77–92). Bloomington: Indiana University Press.

Alley, K. D. (2000). Idioms of degeneracy. In L. E. Nelson (Ed.), *Purifying the earthly body of god: Religion and ecology in Hindu India* (pp. 297–330). New Delhi: D.K Printworld (P) Ltd.

Apffel-Marglin, F., & Paranjuli, P. (2000). Sacred grove and ecology: Ritual and science. In C. K. Chapple & M. E. Tucker (Eds.), *Hinduism and ecology: The intersection of earth, sky, and water* (pp. 291–316). New Delhi: Oxford University Press.

Baindur, M. (2009). Earthed-experiences of nature: Place and nature in Indian Vedic and Purānic thought. *Environmental Philosophy, 6*(2), 43–58.

Baindur, M. (2010a). *Conceptualization of nature: Towards a philosophy of conservation and action from Indian traditional thought*. Unpublished Ph.D. dissertation, Manipal University.

Bilimoria, P. (1997). *Environmental ethics of indian religious traditions*. Paper Presented at the Symposium on Religion And Ecology during American Academy of Religion Annual Conference, San Francisco, November (Unabridged essay copy, Courtesy of the author).

Bowie, F. (2006). Religion, culture and environment. In the anthropology of religion: An introduction (2nd ed. 2000, pp. 107–133). Malden: Blackwell.

Casey, E. S. (1993). Getting back into place: Toward a renewed understanding of the place-world. Bloomington: Indiana University Press.

Chapple, C. K. (2000). Towards an indigenous Indian environmentalism. In L. E. Nelson (Ed.), *Purifying the earthly body of god: Religion and ecology in Hindu India* (pp. 13–37). New Delhi: D.K Printworld (P) Ltd.

Chowdhury, A. R. (2000). Hinduism. In H. Jean and J. Bowker (Eds.), *Sacred place*. London: The Continuum International Publishing Group Ltd.

Das, K. (1964). *Temples of Tamilnadu* (3rd ed.). Mumbai: Bharatiya Vidya Bhavan.

Eck, D. L. (1981). India's 'Tīrthas' : ' Crossings' in Sacred Geography, *History of Religions* (The University Of Chicago Press), 20(4), 323–344.

Eck, D. L. (1990). *Banaras: City of light*. Princeton: Princeton University Press.

Eliade, M. (1959). *The sacred and the profane: The nature of religion*. New York: Harcourt, Brace.

Flood, G. D. (1993). *Introduction in mapping invisible worlds* (pp. 1–5). Edinburgh: Edinburgh University Press.

Gottelieb, R. S. (Ed.). (2004). *Introduction in this sacred earth: Religion, nature, environment*. London: Routledge.

Griffith, R. T., & Hotchkin, H. (1973). *The hymns of Rig Veda*. New Delhi: Motilal Banarasidass Publishers.

Jacobsen, K. A. (1993). Restrained persons, places transformed: Relation to place in the Saṃkhya-Yoga Traditions. In R. B. Singh (Ed.), *Environmental ethics: Discourses and cultural traditions, a festschrift to Arne Naess*. Also published in The national geographical journal of India, 39(1–2): 141–49.

Kane, P. V. (1973). *History of the Dharmaśāstras* (Vol. 4). Poona: Bhandarkar Oriental Research Institute.

Keith, A. B. (1925). *The religion and philosophy of the Veda and Upanishads: Part I*. New Delhi: Motilal Banarsidass Publishers.

Kinsley, D. (1995). *Ecology and religion: Ecological spirituality in a cross cultural perspective*. New Jersey: Prentice hall.

Kinsley, D. (2000). Learning the story of the land: Reflections on the liberating power of geography and pilgrimage in the Hindu Tradition. In L. E. Nelson (Ed.), *Purifying the earthly body of god: Religion and ecology in Hindu India* (225–246). New Delhi: D.K Printworld (P) Ltd.

Mathews, F. (1994). *The ecological self*. London: Routledge.

Merchant, C. (1980). *The death of nature*. NewYork: Harper, SanFrancisco.

Nagarajan, V. (2000a). The Earth as a Goddess BhūDevī: Toward a theory of 'embedded ecologies' in folk Hinduism. In L. E. Nelson (Ed.), *Purifying the earthly body of god: Religion and ecology in Hindu India* (pp. 269–296). New Delhi: D.K Printworld (P) Ltd.
Nagarajan, V. (2000b). Rituals of embedded ecologies: Drawing Kolams, marrying trees, and generating auspiciousness. In C. K Chapple & M. Evelyn Tucker (Eds.), *Hinduism and ecology: The intersection of earth, sky, and water* (pp. 453–468). New Delhi: Oxford University Press.
Rappaport, R. A. (1979). *Ecology, meaning and religion*. Berkley, CA: North Atlantic Books.
Ruether, R. R. (2005). *Integrating ecofeminism, globalizations, and world religions*. Oxford: Rowman and LittleField Publishers Inc.
Tanaka, H. (1988). On the geographic study of pilgrimage places. In S. M. Bhardwaj & G. Rinschede (Eds.), *Pilgrimage in world religions* (pp. 21–40). Berlin: Reimer Verlag.
Thapar, R. (2004). *The setting in somanatha: The many voices of history*. New Delhi: Penguin Books.
Tuan, Y. F. (1974). *Topophilia: A study of environmental perceptions, attitudes and values*. New Jersey: Prentice-Hall Inc.

Chapter 8
Nature in Vedic Thought: Gods, the Earth, and *Ṛta*

Abstract Continuing with the theme of sacred nature, this chapter traces sources of some of the ideas about nature and the human being from the Vedic tradition. The representation of "nature and human" in the Vedic period can be viewed from two perspectives. The representation, thought, and behaviour of the people during the Vedic period from an eco-sociological viewpoint differ from the eco-philosophical interpretation. The purpose of this chapter was to look beyond the two rather opposite viewpoints taken by previous research in this area to see whether it is possible to construct an ecological philosophy of Vedic period fairly, without bias. The term *ṛta*, an alternative conceptualisation of nature as "natural law", and the interpretations of these concepts are also detailed in this chapter. I have also briefly suggested the ecological implications of such understandings.

Keywords Vedic nature · Gods · Goddesses · Panentheism · Earth · Natural law · *Ṛta*

8.1 Sacred Nature: Gods and Goddesses of Vedic Cosmology

One of the dominating influences on Indian thought has been the body of the literature known as the Vedas, and subsidiary texts called the *purāṇas*. These texts are based on a complex understanding of the cosmos and its structure. They construct a story of a universe in which human beings occupy a place within the web of creation and are not separate from it. From the religious and philosophical constructs found in the texts of the Vedic period, one can posit with reasonable certitude that the people who composed the Vedas subscribed to a cosmogony that was not confined to the visible physical environment that they lived in. Space and time were a part of this created cosmos, as were the sun, moon, and the stars. One could say that that it is possible to look at nature as a vision of an interconnected "cosmos" in the Vedas.

M. Baindur, *Nature in Indian Philosophy and Cultural Traditions*,
Sophia Studies in Cross-cultural Philosophy of Traditions and Cultures 12,
DOI 10.1007/978-81-322-2358-0_8

The cosmogonies or the stories of the universe in these texts describe not only our world that is presented to our senses but also mythical worlds of other beings. Within this world view, the cosmos consists of many worlds called *loka*s (which are sometimes mythical and not *on* the earth) and also beings (*bhūta*s) that inhabit such worlds. Vedic verses often refer to the cosmic being—*Vaiśvānara* or *Hiraṇyagarbha*.[1] The various components of this universal being-body are the waters, sun, moon, and the earth. Like the limbs of the body, the various components of the whole cosmic being make up this cosmic realm.[2] The *Puruṣa Sūkta* (*Ṛg Veda* 10.90) for instance, describes this cosmic being as having a thousand limbs and heads and suggests that he permeated the universe and all his creation in an act of sacrifice (Monier Williams 1876, p. 240).

It is interesting to note that these metaphysical and experiential components, like the nature gods in the Vedas are amorphous, lending themselves to multi-level interpretation. The *loka*s in traditional myths are named after the beings that inhabit them. Just as the beings are "enworlded" by the *loka*s, the *loka*s are populated or "en-being-ed" by the inhabitants. The inhabitants often define the place-world (*loka*) by giving them a particular name: *Nāgaloka*: the world of *Nāga*s, snake people, *Devaloka*: the world of *Devas*, divine beings, etc. Sometimes the stories call these worlds by the name of the primary deity whose presence is primary (some other word?) in that world. The trinity gods in Hindu thought are assigned special *loka*s such as *Śiva-loka*, *Viśṇu-loka,* and *Brahma-loka*. The seven earlier hierarchical worlds described in Vedic texts give way in later Puraṇic stories, to the more personal *loka*s in the narratives which are "inhabitant-centric or deity-centric worlds". While these worlds have spatio-temporal dimensions, their material or geographical locations are uncertain. Some worlds are "attainable" worlds, which one reaches by designated rites or austerities, and yet others can be reached by travel and sometimes within one's own mind through meditation. The world view about the *loka*s remains constant; however, retaining the cosmography of the hierarchical planes, with the upper worlds, having more sanctity than the so-called lower worlds. As discussed in Chap. 7, the important point is that the cosmic order within these is maintained and balanced harmoniously by the natural order called *ṛta* (=rita) or *dharma* in various texts (Keith 1925).

The moral hierarchy also determined the sacred precedence of the various beings; a sacred being was defined by the actions and rituals of who salutes whom or reveres whom, who is given offerings first or who is morally in perfect integrity. The origins, the actions of dharma, and adherence to universal laws, all formed a complex indication of the place order of all objects and beings in the cosmos. This cosmology forms an important presupposition for the analysis of the ecologically relevant themes in Indian thought.

---

[1]*Vaiśvānara* is the form of the universal experiencer, consumer. *Hiraṇyagarbha* refers to the cosmic womb.

[2]Also sometimes called *Brahmāṇda* or the cosmic egg.

On examining the Vedic literature and related texts, it is clear that Vedic people worshipped parts of nature as sacred and powerful cosmological entities. MacCulloch (1994) suggests that people living close to nature are stirred by a sense of wonder and awe, which manifest as fear, love, or respect. Perhaps this was the case with the Vedic people who deified the sun, moon, and stars and all that they saw around them (pp. 201–202). This seems to be a rather simplistic reading of the Vedas, which are not merely poems of praise of the powerful elements and natural climatic events.

About the divinities themselves, Jamison and Witzel remark that there was no particular way in which deities could be clearly categorised.

> Indeed, what is striking about the Vedic pantheon is its lack of overarching organisation. Some gods are transparently "natural"—their names merely common nouns, with little or no characterisation or action beyond their "natural" appearance and behaviour (e.g., *Våta,* [*Vāta,*], deified 'Wind'). Others are deified abstractions, again with little character beyond the nouns that name them (e.g., Bhaga–'Portion'). Others belong especially to the ethical and conceptual sphere (e.g., *Varuna, Mitra*), others to ritual practice (Soma, the deified libation) (Jamison and Witzel 1992).

However, as we have seen earlier, the idea of sacrality, even among the Vedic deities, was strictly determined by a cosmic hierarchy of different *lokas*, beings, and moral precedence.[3] Prevalent literature does take on a romantic view of the idea of Vedic natural gods, suggesting that we can somehow learn or derive an ecological sense from these beliefs about sacred nature divinities. The tendency to be anecdotal and situate the verses out of the larger framework of the Vedas to prove a point that the Vedic people were more mindful of the earth than we are seems to be a dubitable discourse.

As I mentioned earlier, hymns of praise in the Vedas are not limited to the earth alone. The sun, the waters, and Indra are praised in turn and asked to protect mankind and issue favours. While I agree that we are no longer in the age where we continue to practice all the Vedic beliefs, it is also impossible to ignore the larger context in which these gods were given offerings. If we examine the philosophy and the belief system behind the praise of these various gods including the earth, we find that the Vedic people had a give-and-take relationship with the universe or cosmos. A review of the philosophy behind rituals and myths shows that the people lived by a rule of reciprocity (Mylius 1973):

> Perhaps the most obvious of the motivating ideas of Vedic religion is the Roman principle of "*do ut d.s*"[sic], "I give so that you will give" (Van der Leeuw 1920–21), or in Vedic terms: "give me, I give you", *dehi me dadāmi te,* TS 1.8.4.1, VS 3.50 (p. 476) (Mylius Quoted by Jamison and Witzel 1992).

This reciprocity extended beyond the community to the phenomenal world, in a system that Jamison refers to in his paper as "natural economy". According to this

[3]Some dysfunctional aspects of this moral order could be the social domination of Brahmins or Kṣatriyas who are given precedence over the others in the community and access to sacred objects, and study of scriptures. Though interesting, this discussion is out of the purview of this book.

view, all nature participates in a cycle where nothing is wasted or lost. For instance, heavenly water falls as rain to earth, it produces plants (which are eaten by animals); both plant and animal products are offered at the ritual and thus ascend to heaven in the smoke of the offering fire, to become rain again. Some philosophers, however, do not believe that these rituals had any deeper meanings than performance.

The contention of most scholars who studied the religious background of the Vedic people suggests that Nature was considered powerful, the human was indebted to these nature gods, whether they were abstract or primitive. It is a philosophy of exchange rather a deep sense of love, respect, or care for nature per se. Writers such as Dwivedi (2000) tend to articulate this as a simple ethics of care when using words such as "*Dharmic* ecology" to refer to the relevance of sacred texts in raising ecological consciousness. We also find that certain readings of these texts assert that the sentiments in the *Pṛthvī Sūkta* denote the bond between the earth and human beings and exemplify the relationship between the earth and all living beings, humans, as well as other forms of life. The interest in classic texts such as the Vedas and *purāṇa*s is important in the search for conceptual resources that may provide answers to issues in environmental ethics (Callicot and Ames 1989).

Patton (2000), however, warns us about reading ecological ethics into every text of the Vedic period. She accepts in her article that affirmation of harmony between humans and nature could add to our constructive behaviour, yet cautions that deliberate readings that leave something out of the ancient assessment of nature would damage the ecological cause. Transparency and critical assessments of such texts, she suggests, may have a better persuasive power. One must be aware that it is not necessary that every sacred or traditional reference to nature would automatically entail ecologically relevant behaviour. Narayanan (1997) also wonders at the disparity between the world views on nature and environmental behaviour and acknowledges the competing forces within the traditions. Given the fact that the Vedic sacrifices and the direct worship of these nature deities is no longer predominant in the popular imagination of the people, it is not possible to re-invoke nature gods within the larger traditions in society. However, it is possible to explore the spirit and the virtues extolled as a context for the Vedic fire-ritual activity—the concepts of reciprocity, offering and also greater good.[4] Patton (2007, p. 127) interpreting the Vedic rituals from a Levinasian framework of ethics suggests that the Vedic "mutuality occurs on a two-folded level". The first of these is between the divinity and the humanity and the second between each of the priests who are active performers of the fire ritual. She suggests "… such a sacrifice involves face-to-face interaction, both at the level of performance as well as the symbolic language of interaction between humans and gods" (pp. 127–128). She refers to the various ways in which a face-to-face accountability can be envisioned in the acts of the fire

[4]In Sanskrit texts, these would be *dharma, dāna,* and *loka-kalyāna,* often discussed together in the *Bhagavadgīta*. See Bilimoria (2007) for an overview analysis of various ethical streams in Indian thought.

ritual (p. 131). As Patton admits, the ethical imagery in the Vedas is far more complex because of its intermeshing with the social, "yet it also carries the possibility carrying the other within oneself" (pp. 141–142).

8.2 Dematerialisation of the Vedic Gods: Loss of an Ecological Narrative

Narayanan (2001) refers to the sanctity of the earth, the rivers, and the mountains in Hindu sacred texts starting with the Vedas. An important historical turn in the conceptualisation of Vedic gods is the dematerialisation of the so-called "natural gods". As discussed earlier, there is no documentation as to why some deities were abstract, some very physical or why some had multiple names and personalities. Most gods associated with real physical aspects of the environment seemed to lose their embodiment in nature and become abstract or a "controlling power".

Social ecologists such as Gadgil and Guha (1992) also interpret the loss of sacred geography to the creation of abstract nature gods such as *Varuṇa* or *Agni* (Brahmin gods). They argue that divinity individualised in trees, groves, and ponds was replaced by abstract forces of nature such as wind, water, fire, air, earth, etc., and these powers were again used to subjugate hunter–gatherers. Geden (1926), however, prefers to consider the transformation of the spirits of nature—trees, ponds, and rivers into a later tradition of deeming certain places and trees, plants, or animals as sacred.

The word for nature used in common parlance in languages derived from Sanskrit, *prakṛti, also* comes to us from the literature of the Vedas and *purāṇas.*[5] Jacobsen (2002) also mentions that the word *prakṛti* is used as a technical term in the Vedas. With reference to rituals, *prakṛti*—means model or archetypal sacrifices. Though there was no particular concept for nature in the Vedic age, the word *prakṛti* has come to represent nature for us in the current times and its conceptual meanings are various, as we have seen in the previous chapters.

8.3 Nature in the Upaniṣads

The Upaniṣads are some of the oldest texts in Indian intellectual traditions. A part of the Vedic literature, the Upaniṣads are distinguished from the rest of the Veda sections because they emphasise the path of knowledge called *jñanakānda,* as opposed to the rest of the Vedic sections that focus on ritual or *karmakānda* (Dasgupta 2004).

[5]*Prākṛt*, the common language, is so named due its natural formation, as opposed to a well-formed created language, *Samskṛt.*

The term *upaniṣad* is formed from the root word *ṣad* that means "to sit" and the two prefixes, *ni* that means "down" and *upa* which means "near". The term therefore refers to the act of "sitting down near" or in another sense it also means "to draw close to". The word represents both the context and the content of the Upaniṣads. The context is that of a close conversation between the teacher and the taught and the content of teaching of these texts is the revelation of a secret knowledge that leads the student closer to a supreme spiritual state. The Upaniṣads are also called *Vedānta,* a culmination of Vedic enquiry into the nature of the truth (Dasgupta 2004).

Though the central theme of the Upaniṣads is seen as "the hidden self or inner being" referred to as *ātman* (Ganeri 2007), we find a number of references to the lived world and its descriptions that are woven into deeply philosophical narratives. As Sharma (2003) rightly remarks, the usual reading of some of these texts is largely limited to viewing them through the perspective of Advaita philosophy. Such a perspective limits the possibilities of the phenomenological descriptions of nature conveyed in many passages (p. 53). He writes "Such a metaphysics at this stage [during the *Upaniṣads* era] did not entail a denial of the world and nature, at least in the same way, as later came to be associated with the school of *Advaita Vedanta*". (p. 52). Sinha (2006) suggests, "Monotheism in the *Veda-s* [Vedas] led to monism" [non-dualism] (p. 4). Therefore, it is equally possible that the Upaniṣads propound diverse streams of thought that represent both dual and non-dual understandings.

Culp (2009) citing Whittemore (1988, 33, pp. 41–44) writes

> Although there are texts referring to Brahman as contracted and identical to Brahman, other texts speak of Brahman as expanded. In these texts, the perfect includes and surpasses the total of imperfect things as an appropriation of the imperfect. Although not the dominant interpretation of the *Upanishads* [*Upaniṣads*], multiple intimations of panentheism are present in the *Upanishads* [*Upaniṣads*].

Bṛhadāraṇyaka Upaniṣad (henceforth, *BU*) and the *Chāndogya Upaniṣad* (henceforth, *CU*) contain some accounts of nature that are relevant to us as many passages. These Upaniṣads describe the interdependencies of the created beings in the cosmos. Clear examples and indications of sacred immanence are also to be found in many passages such as this verse in the *CU:*

> That deity willed, "Well! Let me entering into these three deities, [fire, water, and earth] through this living-self, differentiate and manifest names and forms." (6.3.2, trans. Gambirananda 1983)

In the *Śvetaśvatara Upaniṣad,* the second chapter concludes that the luminous cosmic self is in fire, water, plants, etc., pervading the whole world. Nature and the entire universe are also conceived as representing the entire universe and its functions as the parts of a single being. This being is not always in the form of a human being, but the cosmos is sometimes likened to a tree, or a horse as in this verse in the *BU*:

> Verily the dawn is the head of the horse which is fit for sacrifice, the sun its eye, the wind its breath, the mouth the *Vaisvānara* fire [digestive fire], the year [is] the body of the sacrificial

> horse. Heaven is the back, the sky the belly, the earth the chest, the quarters the two sides, the intermediate quarters the ribs, the members the seasons, the joints the months and half-months, the feet days and nights, the bones the stars, the flesh the clouds. The half-digested food is the sand, the rivers the bowels, the liver and the lungs the mountains, the hairs the herbs and trees. As the sun rises, it is the forepart, as it sets, the hind part of the horse. When the horse shakes itself, then it lightens; when it kicks, it thunders; when it makes water, it rains; voice [neighing] is its voice (*BU* 1.1.1, trans. Müller 1879).

As we see from the passage above, one of the ways nature is conceived is as one unified whole "being" constituted by its different parts and functions. The unified being in these texts led to other traditions of philosophies where the texts suggest that a search for the innermost core or this substantive support of all beings becomes central to their doctrine. Another way of seeking unity of all nature is to describe layers of essences/substances, each giving rise to another, supporting each other. The *CU* for instance describes this idea: "The essence of all these beings is the earth. The essence of earth is water. The essence of water is vegetation. The essence of vegetation is the man. The essence of man is speech…" (1.1.2, trans. Swahananda 1956). Also in the *Muṇḍakopaniṣad,* we find a verse (2.1.9, trans. Gambhirananda 1989) that describes the world of creation as arising from one who chooses to remain as the inner being of all. In the list of beings that have arisen out of this self-being are "all the oceans, all the mountains, rivers, plants, and their juices".

Many verses of these texts also celebrate this inner essence as life or *prāna.* The conceptualisation of a unitary form of life force in all beings of is an interesting view of an organic nature. Some passages for instance mention the presence of life in trees as similar to that of human beings.

> Of this large tree, dear boy, if anyone were to strike at the root, it would extrude sap, though still living, if anyone were to strike in the middle, it would extrude sap, though still living, if anyone were to strike at the top, it would extrude sap, though still living. As that tree is pervaded by the living self, it stands firm drinking constantly and rejoicing (*CU* 6.11.1, trans. Swahananda 1956).

8.4 On the Sacred Earth

One of the oldest conceptualisations of the idea of the natural world is that of the earth. The earth was often considered to be the womb of all living beings. Among the very many Vedic deities associated with nature and natural phenomenon, the closest to our concept of nature is Earth or *Pṛithvī.* As the great mother of all beings and a goddess, the reverence for the earth is unmistakable. In Vedic literature, the earth is regarded as a divine mother. Many writers have pointed out that such a conceptualisation of Earth as a divine mother demonstrates the reverence that the composers of the Vedas had for the land they lived on (see Kinsley 1998; Gottlieb 2004).

Patil (1974) traces the conceptualisation of the earth to a pre-Vedic period. He suggests that the goddess Nirṛti mentioned in the Vedas is a pre-Aryan fertility mother goddess who continues to be worshipped in different forms even in later periods and is mentioned in various texts such as the *Mārkendeya Purāṇa* (p. 36) and the *Vajaseneyi Saṃhitā* (p. 37). The goddess is represented in two aspects in the *Taittarīya Saṃhita* (*TS*): as a cruel deity and a death goddess she is terrible and she is also the Earth:

> Thee in whose cruel mouth here I make the offering,
> For the loosening of the bonds,
> As "earth" people[6] know thee,
> As *Nirṛti*, I know thee on every side. (*TS* IV.2.5 trans. adapted from Patil 1974, p. 31)

He traces the origins of this goddess to the ancient deity who might have been a river goddess: "*Ṛti* or *Ṛta* means, according to *Nigṇṭu* (I.12.68), water. This etymology denotes *Nirṛti* was originally an *apsara,* a water or river goddess" (p. 33).

Patil (1974) also makes an observation that the goddess, Nirṛti, is connected to the realm of water, serpents, tribes, and the land. Using textual quotations and also observations of practices related to this fertility goddess, he establishes the connections between waters in the context of both rivers and Varuṇa, the deity of serpents, and water. Prakṛti, the universal mother goddess (the later creatrix of the Sāṃkhya philosophy) may have been derived from the concept of *nirṛti,* he suggests. If the meaning of *ṛta* is taken to mean cosmic law, this deity becomes the power of those laws, a precursor to *prakṛti.* Patil (1974) concludes his study with the suggestion that "The three characteristics of *Nirṛti,* over lordship over water, land and tribe, in course of time were transformed into the three *guṇas*" (p. 55). Each of the colours associated with the three characteristics, white with water, black with the earth, and red with the over-lordship of the tribe, give rise to the colours associated with the three *guṇas* of *prakṛti.* In some passages in the *TS,* the goddess Aditi is also identified with the Earth, and in others, she is mentioned separately along with the earth. The *Naighantaka* names Aditi as a synonym of Earth—*pṛthivī,* cow—*go,* and in the dual as similar to the heaven–earth pair—*dyāvāpṛthivī* (Macdonell 1998, p. 121). Aditi is referred to as a personification of a universal or cosmic nature. It is interesting to note that the two qualities of Aditi described as prominent are that of motherhood and that of a power to release the bonds of human suffering. While motherhood is closely linked to the idea of the earth, the second power is represented in the cow (Macdonell 1998, p. 121). He suggests that etymologically, as Aditi, it is possible that she is a cow, representing boundless plenty: "Mystical speculation on the name would lead to her being styled a cow, as representing boundless plenty, or to her being identified with the boundless earth heaven or universe" (p. 121).

We find that in the Vedas, there are two forms of reference to the earth. Sometimes, the earth is called earthly plane (*bhūloka*), the *metaphysical* realm of which is a part of either a seven world system (*saptaloka*), or sometimes a three-world system

[6]Patil's (1974) translation reads 'men' for *janā*; I prefer to use the term 'people' here.

consisting of the earthly plane (*bhū)*, the intermediate plane (*bhūvah*), and the heavenly plane (*suvah*) (Baindur 2010).[7] Another form of common reference to the Earth is the land that on which we live on, the earth that provides all the natural resources for all beings, often called the "Goddess Earth". This idea of the Earth revered as a mother is evident in many verses of the Vedas such as the *Pṛthvī Sūkta*. It is assumed that an attitude of reverence would lead to ecologically responsible behaviour towards the planet.

An eco-feminist's critique of this view would dismiss the metaphor as something that is based on ecological romanticism, which chooses to ignore social realities about nature and women. This metaphor is problematic as it still is within the framework of a patriarchal imagination of a vulnerable nature that is to be "protected" (Roach 1996).

In early references in the *Ṛg Veda* for instance, the earth is portrayed as a powerful goddess who is mighty and sustains all beings and landscape features such as mountains (Dwivedi 2000). Transformation of the social life and position of woman in stratified communities led to devaluation of the earth. In an earlier paper, I trace the conversion of the Goddess Earth into a divine abstract, Bhūdevi, and the material resource earth (Baindur 2010). Mani (1989) suggests that the separation between the earth and its body is evident when the texts suggest that "The earth is made of mud and *Bhūdevi* is its Goddess". This reduces the corporeal earth into a resource while the divine qualities of the Vedic Earth are enshrined in the divine consort of Viṣṇu. The Earth goddess loses her power and autonomy and as mere land becomes subservient to the owner—the king or the landlord. The dematerialisation of the earth also occurs with time. The earlier verses clearly indicate that the Earth was venerated as the earth that was material, composed of mud, dirt, and supporting mountains and trees. For instance, the verse in AV XII.1 says:

> Earth is composed of rock, of stone, of dust;
> Earth is compactly held, consolidated.
> I venerate this mighty earth the golden breasted (trans. quoted from Panikkar 1983 p. 125).

Like the West, a scientific narrative of a mechanical universe (Merchant 1990) did not replace the concept of a living earth in Indian traditions. Instead, the objectification was due to cultural and religious norms as well as myths around the roles and characteristics of a woman that seem to augment the patriarchal discourse. A feminist critique of this position would reject the metaphor of the earth as drawing on the romantic idealisation of "woman as mother" and an idea of chivalrous protection of woman, both of which have patriarchal underpinnings.

Drawing from her conclusions on Vedic and Purānic literature, I discuss the problems and possibilities with the two models of belief (Baindur 2010). Firstly, in Indian thought, the earth has been accorded a divine status and is revered as a goddess. Given the reverence shown to the Earth in Indian culture, the unsustainable extraction of resources from the land is a contradiction. The problem with

[7]This study related to this topic has already been published independently as a paper and is to be a separate chapter that is to be published in an edited volume. I briefly summarise my work here.

deification is that any discourse about the Earth as a goddess who is divine and therefore indestructible often masks exploitation and neglects towards the corporeal earth. Secondly, the earth is often regarded as a mother of all beings. As a mother, as land, and as a divine goddess—all these representations seem to form beliefs that could influence people to pro-environmental behaviour. Despite all these beliefs, forms of extracting resources from the Earth are justified by narratives that can be traced to the historical conceptualisation of the Earth and woman in Vedic and Purānic thought. In my study, a critique of this idea is developed through analysis of narratives about the earth in both classical and popular literature linked to the problems with the image of the feminine mother and also of a goddess in a patriarchal society.

The concept of the earth within the Vedic narratives gradually underwent a transformation that was influenced by the sociocultural transformations. It is clear from various textual sources that the earth which is conceptualised as a mighty mother is transformed into a suppliant goddess, and at the same time, she is also distanced from the actual land which becomes de-sacralised and non-organic. I posit that this occurs through narratives and myths such as that of King Pṛtha that legitimise the exploitation of resources.[8] Following my analysis, one can see how the earth, though not called Aditi in the myth of King Pṛtha, comes to embody these two forms of the mother and the cow (Baindur 2010).

I have suggested in this context that it is possible to rework this image of the Earth–mother in a constructive way. The Earth is vulnerable to human action, but this vulnerability need not be connected to the idea of the woman. Instead, if we seek to humanise the earth, giving the idea a temporal context, we can say that the vulnerability of the earth is linked to ageing. In terms of how much longer the Earth can care for us, we have an option of re-imagining the earth as an aged parent who deserves our care and love.

I also point out that perhaps, the humanisation of the earth as an aged parent can restore its fragility and evoke an ethics of care towards it:

> Inside this view, we would need to re-embody the earth having a temporal existence. The solution thus lies not in revering the earth as divine but humanizing it. This explicit invocation of temporality into both humans as children and earth as mother is the first ethical move in a narrative of the earth. Basically this implies that the earth, which was earlier deified, has to be humanized for any ethics to be possible. The metaphor of an unchanging divine and ever young earth that is divine and untouched by our activities has to be replaced with a narrative that foregrounds the earth as an aging mother (Baindur 2010, p. 581).

[8]According to one version of the story (Mani 1989; Baindur 2010), the Goddess Earth had withdrawn all her vegetation into herself and people were suffering for want of food crops. King Pṛtha who was angered by this behaviour of the earth went after her to punish her. The goddess took the form of a cow and ran to all the worlds (*lokas*), but found no place to hide. The goddess was forced to surrender to King Pṛtha and the threat of his powerful bow and sharp arrows. She was then milked for all her resources.

8.5 *Ṛta* and *Dharma*: The Natural Moral Law and Duties

The idea of *dharma* is still based on the concept of *ṛta*, founded on the hedonistic principle that finds mention in the Vedas. This is the principle that guides and ensures performance of one's *karma* (action) according to one's *dharma* and leads to the reduction of suffering; the non-performance of one's *dharma* leads to suffering. As *ṛta* as described in the Vedas, it is not merely an automatic "natural" law that governed planetary movements and the duties of the various gods. In the Indian conceptualisation, *ṛta* evolved into a principle of moral action and righteousness, its meaning taking on similarity with "truth". The word for truth as "integrity" or righteousness is *ṛtam. Ṛtam* is a principle that goes beyond the descriptive truth we call "honesty". Instead, it represents a state of moral affairs, a statement that is a creative act of keeping one's word or what in an Indian perspective would be called as "obeying one's socio-cosmic and place-order duties or *dharma*". Bilimoria (1997) explains that the concept of *ṛta* is important as it connects the created cosmos to a moral order within it:

> The highest good (summum bonnum), however, expresses itself in total harmony of the cosmic or natural order characterised by *ṛta*: this is the *telos*, the creative purpose that underpins human behaviour. The prescribed pattern of social and model and moral order is thus conceived as a correlate of a natural order (p. 33).

The idea of *ṛta* is important in the conceptualisation of nature as the idea of the moral or ethical in Indian thought called *dharma* arises from the same conceptualisation and the word. A discussion on the meanings and various references to *ṛta* would be a merely descriptive project unconnected to the central idea of the philosophy of nature. Therefore, the idea of *ṛta* in relation to its derivative *dharma* becomes a study of natural law in Indian philosophy.

We have earlier discussed the concept of *puruṣartha*s with a view to understand the difference between human and non-human beings of creation. Here, we shall see a moral framework for environmental ethics is possible within the framework of *dharma*. Without attempting to translate the word, which lacks an equivalent term in the English language, we can explore the cluster of ideas around this particular concept which has many moral connotations in Indian thought.

The idea of *dharma* comes from the root *dhṛ* (which means sustaining) and the word *ṛta* that was used in the Vedas to represent moral order or natural law. On one hand, the word *dharma* represents some sort of natural, summing each character of function in any existent. It is commonly used in sentences such as "it is the *dharma* of fire to burn" and again "it is the *dharma* of water to flow downwards". The same word is used in many moral and social contexts too. It represents the norms and the duties enjoined by one's place in the universe. This place order may be cosmic or social or both. According to this, it is the *dharma* of the sun to travel across the sky (cosmic duty) and the *dharma* of the son to obey his father (relationship duty). It is also the *dharma* of a *kṣatriya* to fight (caste/community duty) and the *dharma* of the king to protect his subjects (social/political duty). It is the *dharma* of every human being to tell the truth and keep his promises (a common human moral duty).

As the prescribed and proper function or action, *dharma* can be a moral concept. It is true that the combination of both the natural (in some cases, the identity criteria) and the social context for *dharma* creates a framework in which actions are natural, moral, and social at the same time. Bilimoria (2007) describes *dharma* as moving from a natural order that is organic towards a more human-based ethics rights and duties (p. 37). Mohanty (2007) points out that classically *dharma* is defined by various philosophies in different ways, such as this functional definition of serving a purpose of good: "'*Dharma*' is that from which well-being (*abhyudaya*) and the highest good (*niḥśreyasa*) come about". He also adds

> An answer to the question 'What is *dharma*?' which abandons the project of defining it is this: '*dharma* is that which the cultivated persons (*āryāḥ*) praise when it is done, and *adharma* is that which they condemn when it is done' (pp. 59–60).

One of the aspects of *dharma*, referred to as the *varna-āśrama-dharma* relates to the idea of "caste duties" and is unfortunately, the source of much debate in the modern Indian society. This is one of the particular types of *dharma* that may have to be ignored in this discussion and we may to instead consider a broader understanding of dharma as place-order-based duties or obligation for a philosophy of nature. This would also help to interpret *dharma* within a framework of environmental ethics. The idea of *dharma* takes away from the question of moral considerability (as discussed in Chap. 4) and instead focuses on the moral agent and his actions. Rolston (1999) remarks that one of the difficulties of an anthropocentric and personalistic ethics of the Western world is that

> According to holders of the humanistic perspective, humans can have no duties to rocks, rivers, or ecosystems, and almost none to birds or bears; humans have serious duties only to each other, with nature often instrumental in such duties; the environment is the wrong kind of primary target for an ethic; nature is a means, not an end in itself; nothing there counts morally; and nature has no intrinsic value (p. 410).

Rolston (1988) also derives the concept of duty to nature from different values which are instrumental such as life support value, economic value, recreational value, scientific value, aesthetic value, as well historical and cultural values. On the other hand, in Indian thought, duty to nature is given by the internal benefit that the human being derives from following *dharma*. It is an instrumental value but it is not derived by the use of nature directly. The benefit of all actions is twofold. On one level, there is an immediate effect on our everyday existence. On the other hand, we have another level of effect that is a moral judgement. Rather, all the actions give the result moral *karman*. The naturalisation of moral obligations results in nature being instrumental in actually "fulfilling" a human being's need to follow his/her *dharma* rather than "deserving" of his interventions or moral obligations. Nature becomes a means for human beings to attain liberation and gain merits. Nature is instrumental, not in the sense of being exploited, but in the sense of becoming a field for human *karma* and the fulfilment of *dharma*.

Extending this to the modern context of environmental ethics, the idea of *dharma* can be easily applied to our duty to care for the environment, the non-performance of which will ultimately lead to suffering of the human kind. The idea of one's *dharma* is fundamentally based on a being's relationship to the world around her. To act in a "*dharmic*" way is to act according to one's place and context in the universe. So the idea of *dharma* forms a foundation for an ethics that is already based on human relationship to the rest of the beings, cosmos, or even one's own self. This can actually provide a framework where one's ethical or moral actions are based on norms given by the place order state of being. This framework of ethics is certainly anthropocentric, but the foundation of it is based on non-anthropocentric, cosmic view of the human beings' place as part of a larger cosmos. The principle of ecological ethics in Indian thought is fundamentally based on the unique, internally relational, substantive, yet functionally differentiated constituents of the universe. These elements find themselves expressed in alternative discourses of meaning-making of the people, whose interaction with the everyday world is often given by narratives rather than by any understanding of "facts" or "concepts". This world view is combined with a strong normative principle of action, where being and function are interrelated. To be human is to be within the realm of both *ṛta* and *karman* and that means to be related to every other created existent in the world.

8.6 The "Relational", *Dharma,* and an Interpretation of Ecological Ethics

The philosophy of Sāṃkhya in particular upholds that there is an eternal relationship between the whole and its parts, in this case between *prakṛti* and its evolutes. The relationship is a category of *sāmānya* or generality that inheres in all existents.[9] Aniruddha, a Sāṃkhya commentator, is summarised by Larson thus "It is true that materiality [*prakṛti*] and consciousness [*puruṣa*] are eternal, but nevertheless, the category of universal property [*sāmānya*] has a certain constancy".

Vaiśeṣika philosophy gives the definition of the universal *sāmānya* as "eternal one and residing in many". The universal is also explained as the one by the presence of which many individuals belong to a single class.[10] For both the Sāṃkhya and Vaiśeṣika, this is an ontologically real category and is not like a concept in the mind. This makes all universals metaphysical realities that are not mere abstract principles. The universals reside in substances, qualities, and actions and claim the realists. One such relationship universal that inheres in all of *prakṛti* is the universal "*dharma*". The concept *dharma* is a unique universal (*sāmānya*) as it inheres in substance (*guṇa*s or collocations), in qualities as the function of an existent, and also in action as ordained moral action. Further, from the discussion

---

[9]The idea of pursuing the concept of relatedness evolved after a discussion with Dr. Sundar Sarukkai who saw connections within the broad themes I had put forward in an early draft of this chapter.

[10]See the details of *Vaiśeṣika* categories in Sharma (2003).

on *dharma* in chapter three, *dharma* is a unique category in Indian moral thought, because by virtue of being ontologically given, it establishes the relationship of human beings to the cosmos at large. Koller (1972) reiterates that the unique ontological status of *dharma* and *ṛta* is a profound thought of the Vedas:

> But it would be a mistake to infer that since *dharma* and *ṛta* are normative they are not ontological, for they are both. There is no difference between the being (*sat*) of reality and its function (*ṛta*). Just as in the *Upaniṣads*, truth is identified with *dharma*, so in the Vedas, truth (*satya*) is identified with *ṛta* (p. 136).

Connected to the principle of *dharma*, especially in the human realm, is the idea of *karman* or moral action. Koller (1972) emphasises that while *dharma* provides the normative dimension of relatedness, *karman* refers to the "connectedness of events". He adds "The law of *karman* guarantees the relatedness of all events in the world but does not provide for the regulation of events. The ordering or regulation of relations between events is accomplished by *dharma*" (p. 141).

According to Panikkar (1977), *ṛta* and *karman* are not merely mechanical forces; they are functional and relational. He writes "… both *ṛta* and *karman* are always functional and they function according to a set of relational factors, one of which is human will along with its sentiments and feelings". This internal relationship between action, moral, and the universe at large is very significant for a theory of Indian ethics, specifically a foundation for an ethics that is not human-centric. Ethics therefore becomes a process, a "being in a state of relating" to the en-worlded state of human beings. It is clear that the normative aspect of leading a "good life" is analogous to being in state of *dharma,* a state where one performs actions within the context of one's prescribed position in the universe.

The relationship between the substances and their manifestation that is constituted by *guṇas* provides for a framework of evaluation that does away with a categorical view of conservation and replaces it with a relational view. We as human beings, conserve not because we are different and separate from "nature", but because we are also *prakṛti* and relatedness inheres in everything as *dharma.* The same *dharma* inherent in human beings as members of a created cosmos (*nisarga*) embodied in different bodies and en-worlded by *lokas* expresses itself as care towards the environment.

References

Baindur, M. (2010). Humanising the earth: Reversing the 'reverence to resource' in the Indian context, special issue: globalism, transnational, gender and ecological engagements. *Man In India, 90*(4), 567–587 (Guest Editor: P. Bilimoria, with assistance of Amy Rayner).

Bilimoria, P. (1997). *Environmental ethics of Indian religious traditions.* Paper presented at the Symposium on Religion and Ecology during American Academy of Religion Annual Conference, San Francisco, November (Unabridged essay copy, courtesy of the author).

Bilimoria, P. (2007). Introduction to part A: Early Indian ethics–Vedas to the Gītā; dharma, rites to 'right'. In P. Bilimoria, J. Prabhu & R. Sharma (Eds.), *Indian Ethics: Classical traditions and contemporary challenges* (Vol.1, pp. 33–56). New Delhi: Oxford University Press.

Callicott, J. B., & Ames, R. T. (Eds.). (1989). *Introduction. In: Nature in Asian traditions of thought*. Albany: State University of New York Press.
Culp, J. (2009). Panenthesism. In E. N. Zalta (Ed.), *The Stanford encyclopaedia of philosophy* (summer 2009 edition) [online] http://plato.stanford.edu/archives/sum2009/entries/panentheism/. Accessed January 2, 2011.
Dasgupta, S. (2004) (1922). *A history of Indian philosophy*, Vol. 1. New Delhi: Motilal Banarsidass Publishers Private Limited.
Dwivedi, O. P. (2000). Dharmic ecology. In C. K. Chapple & M. E. Tucker (Eds.), *Hinduism and ecology: The intersection of earth, sky, and water* (pp. 3–22). New Delhi: Oxford University Press.
Gadgil, M., & Guha, R. (1992). *Use and abuse of nature incorporating 'this fissured land: An ecological history of India' and ecology and equity*. Reprint, omnibus edition, New Delhi: Oxford University Press (1992) (Page references are to the reprint omnibus edition).
Gambirananda, S. (1983). *Chāndhogya Upaniṣad*, with the commentary of Śankarācārya (trans.). Calcutta: Advaita Press.
Ganeri, J. (2007). *The concealed art of the soul: Theories of self and practices of truth in Indian ethics and epistemology*. Oxford: Oxford University Press.
Geden, A. S. (1994). s.v. Nature-(primitive and savage) Hindu. In J. Hastings (Ed.), *Encyclopaedia of religion and ethics*, Vol. IX. Reprint, (1926). p. 227. Edinburgh: T&T Clark.
Gottlieb, R. S. (Ed.). (2004). Introduction. In *This sacred Earth: Religion, nature, environment*. London: Routledge.
Jacobsen, K. A. (2002). *Prakṛti in Sāṁkhya-Yoga: Material principle, religious experience, ethical implications*. New Delhi: Motilal Banarsidass Publishers, Indian edition.
Jamison, S. W., & Witzel, M. (1992). Vedic Hinduism. [Online] www.ms.uky.edu/~sohum/sanskrit/vedica.pdf. Accessed May 25, 2004.
Keith, A. B. (1925). *The religion and philosophy of the Veda and Upanishads: Part I*. New Delhi: Motilal Banarsidass Publishers.
Kinsley, D. (1998). *Hindu Goddesses: Visions of the divine feminine in Hindu religious traditions*. 1987 reprint. New Delhi: Motilal Banarsidass.
Koller, J. M. (1972). Dharma: An expression of universal order. *Philosophy East and West, 22*(2), 131–144.
Mac Culloch, J. A. (1994) (1926). S.v Nature (primitive and savage). In J. Hastings (Ed.), *Encyclopedia of religion and ethics*, Vol. IX, reprint Edinburg: T&T Clark.
Macdonell, A. A. (1998) (1898). *Vedic mythology*.1995 reprint. New Delhi: Motilal Banrasidass Publishers.
Mani, V (Ed.) (1989). S.v. Bhumi. In *Puranic Encyclopedia,* I Edn. New Delhi: Motilal Banarsidas Publishers.
Merchant, C. (1990). Preface: 1990. In *The death of nature*. New York: Harper SanFrancisco.
Mohanty, J. N. (2007). Dharma, imperatives and tradition: Toward an Indian theory of moral action. In P. Bilimoria, J. Prabhu, & R. Sharma (Eds.), *Indian ethics: Classical traditions and contemporary challenges*, vol. 1 (pp. 57–78). New Delhi: Oxford University Press.
Monier M.-W. (1876). *A Sanskrit-English dictionary: etymologically and philologically arranged with special reference to cognate Indo-European languages*. Revised by E. Leumann, C. Cappeller, et al. (1899), Oxford: Clarendon Press.
Müller, F. M. (trans.) (1879). *The Upanishads*. The Sacred Books of the East: Vol I. London: Macmillan & Co.
Narayanan, V. (1997). One tree is equal to ten sons: Hindu responses to the problems of ecology, population, and consumption. *Journal of the American Academy of Religion, 65*(2), 291–332.
Narayanan, V. (2001). Water, wood, and wisdom: Ecological perspectives from the Hindu traditions. *Daedalus—Journal of the American Academy of Arts and Science*, [online] www.amacad.org/publications/fall2001/narayanan.aspx. Accessed May 25, 2004.
Panikkar, R. (1983) (1977). *The Vedic experience: Mantramanjari*. New Delhi: Motilal Banarsidass Publishers.
Patil, S. (1974). Earth mother. *Social Scientist, 2*(9), 31–58.

Patton, L. L. (2000). Nature romanticism and sacrifice in Ṛgvedic interpretation. In C. K. Chapple, & M. E. Tucker (Eds.), *Hinduism and ecology: The intersection of earth, sky, and water* (pp. 39–58). New Delhi: Oxford University Press.
Patton, L. L. (2007). The fires of strangers: A Levinasian approach to Vedic ethics. In P. Bilimoria, J. Prabhu, & R. Sharma (Eds.), *Indian ethics: Classical traditions and contemporary challenges* (Vol. 1, pp. 117–147). New Delhi: Oxford University Press.
Roach, C. (1996). Loving your mother: On the woman-nature relation. In K. J. Warren (Ed.), *Ecological feminist philosophies* (pp. 52–65). Bloomington: Indiana University Press.
Rolston, H, I. I. I. (1988). *Environmental ethics: Duties to and values in the natural world*. Philadelphia: Temple University Press.
Rolston III, H. (1999). Ethics and the environment. In E. Baker, & M. Richardson (Eds.), *Ethics applied*, 2 Edn. (pp. 407–437). New York: Simon & Schuster.
Sharma, C. (2003). *A critical survey of Indian philosophy*. 2000 reprint. New Delhi: Motilal Banarsidass Publishers.
Sinha, J. (2006). *Indian Philosophy,* vol. 1. (1978) Reprint, New Delhi: Motilal Banarsidass Publishers.

Chapter 9
Nature in Literature: Nature's Presence and Absence

Abstract This chapter will look at some major texts in describing and writing about nature in pre-modern Sanskrit literature such as by Kālidāsa and Bāna. The human–nature interrelationships are constructed through meanings attached to metaphors in Sanskrit poetics and the narrative. Such powerful metaphors construct the imagination of nature from a deeply anthropocentric view of nature. These imaginations also add to issues around conceptualisations of nature as gendered, or as benevolent, or dangerous. The impact of these imaginations that continue to influence our perception of nature today will also be discussed.

9.1 Introduction

Cultural conceptualisations of nature play a significant role in the way human beings relate to the environment in a particular society. There have been many texts in the Sanskrit language that have had an impact on popular imagination, literature, and even performative traditions. Unlike the Romantic Movement in Western thought, some of the nature poetry composed in pre-modern India was not a response to any critical historical event. These earlier poems were written under patronage of kings or they are compositions of epics and stories that were linked to mythology and oral histories. Given the abundance of flora and fauna in the sub-continent and the close interaction between the seasonal cycles and human activities, many of the writers chose to include descriptions of nature and the landscape in their compositions. This chapter breaks away from philosophical themes and forays into cultural traditions of the literature that has influenced many later writers and poets. The sample of writers selected here are significant in their contribution to the cultural ethos of Indian thought. The choice of classical literature discussed in this chapter is a sample of some of the significant texts that are already considered nature related. Two mainstream poet composers of pre-modern Sanskrit literature, Bāna and Kālidāsa, have been selected for their frequent references to nature in some of their texts. Bāna is considered to be one of earliest prose writers, and the

M. Baindur, *Nature in Indian Philosophy and Cultural Traditions*,
Sophia Studies in Cross-cultural Philosophy of Traditions and Cultures 12,
DOI 10.1007/978-81-322-2358-0_9

description of the forests and landscape in his work *Kādambarī* (henceforth, *Kdmb*) has been often praised. This is based on two reasons: one is the easy availability and popularity of the texts of Kālidāsa and Bāna who have been already discussed widely among scholars, making this secondary analysis an easier task. There are also a number of translations and commentaries available. The second reason is to focus on the literature that has been produced at the same time as some of the pre-modern philosophical thought discussed in the rest of the book. This is also not a purely literature analysis but also conceptual. These compositions have also influenced later writers in many other Indian languages, and some of the literary imaginations and themes are used even in the contemporary times.

I foreground the representation of landscape, references to flora, fauna, and the seasons, non-human elements of the environment, and their relation human emotion and activities from the poetry verses and prose passages. Absences and the perspectives of nature in the text are discussed. This does not claim to be a comprehensive study of all texts of this period by any means, but it is hoped that it sets the conceptual foundation for further engagement with nature in the literature and cultural traditions in Indian thought.

9.1.1 Stories and Fables

In Indian literature, we also have parables and stories that have many references to non-human beings particularly the animals. The most popular among them are the *Jātaka* tales and the stories of *Pañcatantra*. The *Jātaka* stories are stories of Buddha's earlier births, including his birth in the non-human world. Older than the *Pañcatantra*, they could have even influenced the animal themes in later stories and narratives. A part of the Pāli canon, particularly in the *Theravāda* school, these stories of *Bodhisattva* form a part of the Buddhist philosophical instruction. These stories, often narrated by the Buddha himself, represent how even as an animal he always showed compassion and virtue. The stories are told in response to a context and end with Buddha pointing out the identity of the reborn disciples or listeners. The versions of the Jātaka find themselves in many different versions of stories and narratives including the *Pañcatantra* and folktales.

In the *Pañcatantra*, said to have been composed by Viṣṇuśarman, the form of the fable is used to narrate stories about animals that exhibit an almost human behaviour. The animals have relationships with each other, make moral and rational choices, cheat, and also betray each other in human ways. These descriptions show that the animals are deeply humanised. While it is easy to dismiss these fables as literature that only utilises the animals to make a point about human morals, one could suggest that there is a particular stereotype of animals that allows the stories to represent particular human-like behaviour. Some common beliefs about animal behaviour are recounted in the stories. There is also categorisation of particular animals as good, bad, stupid clever or even compassionate. There is also a tendency to essentialise the animals into 'noble' or 'wretched' categories. These stereotypes of animals are prevalent in many idiomatic expressions and popular sayings all over India in different forms. One cannot of course assert that these stories created the stereotypes of the animals as the beliefs may have been prevalent in the time the stories were written.

The epics particularly the *Mahābhārata* and the *Rāmāyaṇa* contain many references to the geography of the forests of India. Though, in these texts, the sacred and mythical narratives tend to overshadow a natural history, scholars are continuing to use these texts to understand representations of forests and landscapes within these texts that have influenced dominant culture and traditional practice in India. We have already seen the category of the forest in Chap. 6. The literature of the epics *Rāmāyaṇa* and the *Mahābhārata* also have references to nature as the protagonists of both these narratives spend a large part of their lives exiled in a forest. However, the references to the forest and the transformative role of the landscape and imagination of forest is better dealt with under the general theme of "forests in Indian thought", discussed separately in relationship to landscapes. The epics are a different kind of text and are also linked to sacred geographies. I have used methods of ecocriticism in my conceptual analysis in this chapter. Some of the descriptions of nature in this literature can be classified under a few broad themes to help our analysis of the literature from an ecocritical perspective.

Descriptive features:

- References to environmental conditions and climate
- References to flora and fauna
- Reference to metaphors and simile based on nature and natural objects
- References to human/divine beings interaction with nature
- References to geographical elements such as rivers, hills, lakes and forests

Conceptual features:

- Nature's voice and humanisation of nature
- Nature as an active and as passive
- Imagery and representation
- Absences and inclusive references to non-human
- Perspective and vantage points of narration
- Ecological ethics

We begin with poetry of Kālidāsa that has straightforward references to nature and then analyse two sections of Bāna's magnum opus, *Kādambarī*.

9.2 Kālidāsa: Landscapes and Seasons

Kālidāsa has a number of compositions, both plays and poems, and is perhaps the most famous of the Sanskrit literary giants. Two of his short poems directly relate to the theme of nature and environment. Most of his other compositions also contain references to the natural world, but these two can be considered as belonging to the genre of "nature writing". Commentators and critics alike have pointed this out (Keith 1996; Macdonell 1927). On Kālidāsa's style Kale (1974) has this to say:

> He [Kālidāsa] set forth as a necessary corollary that man attains his true dignity only in realising that is not independent of and above the world that is not human: that the ocean and the rivers, the mountains and the forests, the trees and the flowers, the birds and the beasts are as much conscious of a personal life as man and therefore claim from him a recognition of their dignity and worth. He also notes that the poem intertwines the two themes of nature and human emotions so well that they seem indistinguishable (p. xv).

Kālidāsa is known for his remarkable use of *Upama*, the figure of speech known as simile. Kālidāsa's comparisons are not incomplete or hinted but are elaborated and "drawn out in full" (Keith 1996, p. 105). Metaphor and simile help Kālidāsa mingle the natural and the human worlds, the human beings remind him of nature and nature reminds him of the human beings. In the poem, *Meghadūta* (henceforth, *Mgdt*): for instance, ripples in the Vetrāvati river are compared to the river knitting her brows at the noisy cloud thundering on her banks (*Mgdt*, verse 25).[1] And again —"the eyes of the sorrowing wife are half open and half closed like a lotus bud on a cloudy day neither closed, nor open fully" (verse 30).

Kālidāsa particularly captures in his descriptions all of the five sense experiences. The immersion in the world of nature is not just visual, the touch of the cool sandal paste on the skin, the heady fragrances of the flowers and the sounds of anklets combine to give the reader a complete sensorial satiation. David Abram, the environmental philosopher, suggests that the world of nature can be experienced through the senses. For him, narrative forms a way to reconnect with nature and recreate the relationship with the non-human world. Language can be used as a medium to foster this reconnection:

> It is the practice of spinning stories that have the rhythm and lilt of the local soundscape, tales for the tongue, tales that want to be told, again and again, sliding off the digital screen and slipping off the lettered page to inhabit these coastal forests, those desert canyons, those whispering grasslands and valleys and swamps. Finding phrases that place us in contact with the trembling neck-muscles of a deer holding its antlers high as it swims toward the mainland, or with the ant dragging a scavenged rice-grain through the grasses. Planting words, like seeds, under rocks and fallen logs letting language take root, once again, in the earthen silence of shadow and bone and leaf (Abram 1997, pp. 273–274).

Kālidāsa achieves this in his poetry and immerses the reader in the eros of emotion and the ethos of nature at one and the same time.

[1]Unless mentioned specifically, all verses that are numbered from *Ṛtusamhāra* are from Devdhar (2001) and those from *Meghadūta* are from Kale (1974).

9.2.1 Nature as Seasons: The Ṛtusaṃhāra

The *Ṛtusaṃhāra* (henceforth, *Rtsm*) is regarded as one of the simpler compositions of a younger Kālidāsa. Keith (1996) suggests that perhaps the poet's maturity that we find in his other compositions is missing in this poem. Mallinatha who commented on other poems chose to leave this out. Keith however defends this poem and points out that "The poem is far from a mere description of seasons in their outward aspects, though Kālidāsa exhibits the delicate observation and that loving sympathy with nature which seems innate in Indian poets" (Keith 1996, p. 84). The seasons are a part of the order of nature. Nature and environment change but follow a pattern of an annual cycle to which the humans respond by following seasonal practices—the *ṛtucharya* discussed earlier in Chap. 6. All seasons are seasons of love for Kālidāsa, the season of separated lovers or of those in each other's company. Nature as the flora, fauna, and the water bodies all varies with the seasons, and this is captured by the poet in the *Ṛtusaṃhāra*, the gathering (together) of seasons. The meaning ascribed by translators is the collection of seasons. Kālidāsa does not intend it to be a mere collection. The term *saṃhāra* is used for the verb that describes the "gathering the hair and tying it up or plaiting it". Kālidāsa similarly intends to weave the seasons together as an aesthetic crown on the nature's head. The poet pays attention to all of the senses and tries to recreate for us the experience of the season. Each season is described in a canto and there are six.

Devadar (2001, p. v) in his introduction to the translation of *Rtsm* writes:

> The focus [of the poem] shifts back with ease from the beauty of nature to the charm and wonder of youthful love. The emotional harmony between a particular mood and a particular aspect of the season in question emanates from each piece of description.

The poem begins with summer. Though traditionally the seasons are counted from spring, Kālidāsa wants to end his poem on the joyous ode to spring, the season of new beginnings.

Tandon (2008, p. 33) traces the pattern of each canto and writes:

> Each canto follows a pattern, a design in which various changing aspects of nature and the response of human beings and animals to the same are lovingly delineated.... Every other canto is replete with *Sringāra* rasa, the erotic sentiment of which Kālidāsa was an accomplished poet.

The description begins with the poet drawing attention to water shortage in the ponds due to frequent bathing by the heat tormented people. Ways that people sought to keep themselves cool—by fountains, sandal paste application, or wearing of light clothes—are described in this canto. From verse 1.11 to 1.23 (*Rtsm*), Kālidāsa also describes the effect of the hot summer on animals. The theme is that the heat and thirst are so oppressive that animals behave unnatural. They ignore the predator–prey relationships and ignore the danger to their life, they seek the shadows of their hunters to rest in, and in turn, the predators being thirsty and exhausted themselves do not attack: "Tucking its hood in, the serpent hissing, rests below the peacock" (*Rtsm*, verse 1.13). Similarly, the lion does not attack the elephant and the snake the frog. The deer are fooled by the mirage of the water and wander off. The boars dig themselves into the lake bed, and the birds have flown away to wetter lands. There is description of a forest fire from *Rtsm* verse 1.22 onwards. Kālidāsa brings the heat and the consumption of the flames to the reader's experience:

> Fanned by the wind, the fire rises up in the mountain valleys and crackles and splutters through the (bamboo) reeds. In just a moment it spreads through the grass, gaining speed, tormenting the wild animals in the edges of the forest. (*Rtsm*, verse 1.25, trans. by author)

The description of the rainy season follows the summer. The rains bring cool showers first allay the heat but soon the skies are overcast and the rivers flood their banks. The incessant rain increases the passion of human lovers. The earth is colourful, bedecked with the jewel-like sparkling new vegetation, bright coloured insects, and dancing peacocks. In a very clever verse, the poet seems to rebuke both the rivers and women for trespassing their boundaries: "Like women, in whom wantonness arises, the rivers uproot the families of trees on the banks carelessly, their waters sullied with mud, rushing to meet the ocean" (*Rtsm*, verse 2.7, trans. by author). The poet certainly is not the first to feminise nature; he only reflects the social mores of his times. The comparison of a wanton woman—(probably adulterous) who rushes to meet her lover, her reputation sullied and carelessly destroying families—to a river breaking its banks with families of trees begins to show us a glimmer of Kālidāsa's expertise in humanising nature and skilful use of language. Elsewhere the water streams rushing down the slope look like shiny snakes that frighten the frogs (*Rtsm,* verse 2.13). The fragrance of the flowers *kadamba*, *sarja*, *arjuna*, and *ketaki* wafts everywhere (*Rtsm*, verse 2.17). The rainy season is full of fast movements—dancing peacocks, the rain, the flowing rivers, lightening; clouds are being carried by winds: "The wind fragrant with the scents of flowers moves the branches of the trees, they are swaying like the forest is dancing with joy" (*Rtsm*, verse 2.23).

In the last verse Kālidāsa sums up the cultural reverence for the monsoons in India. He calls the rainy season—"The one who has become the life-force of all the living beings" (*Rtsm*, verse 2.28).

After all the activity of the rains, the autumn season seems calm and almost dignified in its nature. The rivers flow quieter, the skies are white, and clouds are moving slowly across the sky like a procession (*Rtsm*, verse 3.4), and the gentle cooing swans have replaced the dancing peacocks (*Rtsm*, verse 3.11–3.12). The clear nights with the autumnal moon and stars constellations delight all the people but torment the separated lovers (*Rtsm*, verse 3.9).

The flowers of this season are *mālati* and the *shyāma* creeper, the water lotuses. The harvest is yet ripening, being rippled by the wind (*Rtsm*, verse 3.10). We have one of the rare references to agricultural landscape in this verse: Men are delighted by the fields stretched to the horizon that have their ground covered by abundant and healthy rice plants. Herds of cows adorn the landscape resonant with the cries of swans and storks (*Rtsm*, verse 3.10).

The cold season (*Śiśira*) begins with frost and activities are undertaken in the warm sunlight. The plants seem to be wilting and the snow begins to fall. The nights are filled with passionate lovemaking and the days with languorous activities. In the next season of winter (*hemanta*), the mood of poem shifts to accounts of erotic love and the moods of love. The cold is described: "The people do not enjoy the nights which are frozen with the thick fall of snow, and clear star filled skies" (*Rtsm*, verse 5.4). A few verses in between speak of the *krouñcha* birds ripened harvest standing in the fields and the seasonal flowers are some of the only references to nature (see *Rtsm*, verse 5.1).

The season of spring, perhaps the poet's own favourite, is much celebrated. The calls of the *koel* bird and the new blossoms of the mango tree are symbols that are deeply embedded in the cultural descriptions of spring. The red *aśoka* flowers, *kurumba* and *kimśuka* flowers are in bloom. There is a visual profusion of the colour red in these verses. Young red leaves, *palāśa* and the mountain slopes covered with flowers gladden the heart (*Rtsm*, verse 6.25). Kālidāsa ends with the spring season, with a benedictory verse celebrating Kāma, the god of love.

The poetry of the seasons as pointed out earlier is fairly simple, and it is not possible to read any deeper meanings into it. But the celebration of the seasons as imagined by Kālidāsa repeats itself in many cultural texts such as music, dance, performances and paintings. The seasons form an important part of understanding the cyclical nature of seasons and people's responses to it, not only through climatic adaptation but also by evoking moods of each of these *ṛtus*. In the next part, another of Kalidāsa's famous nature-themed poems, *Meghadūta* will be examined.

9.2.2 Kālidāsa's Meghadūta: The Silent Cloud

Kale (1974) in his introduction to *Meghadūta* points out that this is a *khaṇḍakāvya*, i.e. a short poem, where two main themes love and nature are interwoven into the narrative. While it is possible to call the poem as nature poetry, it is also important to note that the poem is human centric and the nature is inclusive of human love and longing.

Each of these categories will be discussed with illustrations from the text. The poem itself is divided into two parts called *Pūrvamegha* (*Pū*) and *Uttaramegha* (*Um*). The plot of the poem is simple and uses the emotion of separated lovers as a central theme. A *yakṣa* (a demigod or divine being) estranged from his wife, talks to a cloud asking it to convey a message to her. The *Pūrvamegha* has extensive descriptions of nature and geography, while *Uttaramegha* has many descriptions of the mythical city Alākapuri, the city of the demigod Kubera, where the home of the *yakṣa* and his beautiful sorrowing wife is situated. The actual message of love is also revealed in this second part.

The context of the poem in the first part is geographical. The *Yakṣa* who is exiled from his home in Alākapuri, on the slopes of mount Kailāśa, is roaming in the ashram near Rāmagiri, among the Vindhya mountains in central India. The rainy season traditionally associated with the meeting of lovers and dalliances, brings much sorrow to the *yakṣa* in exile. He addresses a large cloud and begs it to carry his message to his city Alākapuri. The *yakṣa* then proceeds to describe the route the cloud has to take across the land, the geographical markers along the route and also recommends sites for the cloud to visit as well as places for it to rest. The second part of the poem that describes the city and the mansion ends with the actual message that the cloud has to deliver.

The section of *Pūrvamegha* which has more references to nature is discussed in detail from an ecocritical perspective here followed by a briefer summary of the city and aesthetics of nature within the city. All along the journey the cloud is not treated as a passive witness of the different regions it moves, but the coming of the cloud signifies the beginning of many activities of not only the humans and the non-human worlds, but also the worlds of the divinities. The cloud is also encouraged to interact with the environment, nourishing itself by drinking water, cooling people, putting out forest fires or enhancing the visual aesthetics of the landscape. We could posit that the poet uses this poem to demonstrate the position of the cloud in scheme of things. Does Kālidāsa imbibe the cloud with sentience? An interesting verse right in the early part of the poem dispels any such fanciful notions one might have. Kālidāsa makes it clear that the cloud messenger is sentient only in the mind of the lovelorn *yakṣa*. The verse expresses the incongruity between the insentient cloud and the activity of delivering a message that only living beings may undertake. The poet asks us to excuse the poor *yakṣa* and his state of mind that is incapable of judgement due to extreme emotion of separation from his beloved:

> Where (on one hand) his (i.e. what congruity is there between) a cloud, a mixture of smoke, light, water and wind, and where the import of messages (on the other), fit to (which can only) be conveyed by beings with organs capable of discharging their functions (i.e. men) ! Not at all taking this into consideration the *Yakṣa* addressed a petition to the cloud: for those that are sorely affected by love are naturally incapable of distinguishing between objects, sentient and insentient (*Mgdt* verse *Pū*. 5, trans. Kale 1974).

The cloud is not defined metaphysically in this verse, but described rather as a phenomenon that one would normally dismiss as a collocation of insentient matter that thunders and creates lightening. This is the human everyday experience of the cloud and not what it actually may be. Other descriptions of the cloud are around its dark colour—that it retains throughout the poem—being compared to Lord Kṛṣṇa's colour (*Mgdt*, verse Pū. 15) and the rainbow around the cloud similar to the peacock feather in his hair. The cloud's dark hue is also compared to the well-oiled hair of a maiden (*Mgdt*, verse Pū. 18), and to the dark river Yamuna (*Mgdt*, verse *Pū*. 54), to *kājal* (collyrium/kohl), and to a sapphire (*Mgdt*, verse *Pū*. 48). The reflection of the cloud and its shadow also are described variously. The cloud however tends to change its shape (by will it seems) and loses mass when it rains on slopes or fields. It becomes larger after drinking water from various water bodies. We must note that all these descriptions are given in the *yakṣa*'s voice and narrative.

The *yakṣa* in his condition is desperate enough to address a cloud that is insentient. The poet in the whole poem maintains the silence of the cloud, giving it no voice or response to the *yakṣa* at any point of time. The absence of a voice for the cloud is only an indication that the cloud is not rational and has no senses to listen or a mouth to speak. In the later description, the cloud engages in all the activities that a cloud by nature performs, moves along with wind, rains, thunders and absorbs water (it was believed that clouds absorb water from water bodies). Many scenic descriptions are also given from the cloud's viewpoint. In the first part of the poem, poet decentres the human beings in a way that creates a cloud centric perspective of the world. The view is a "looking-down-on-the-landscape" view. Mountains for the cloud to rest on, vistas of landscape below on the ground, the features of landscapes forming pictures of a beautiful woman's body, water features that refresh the cloud or are enhanced by rain, and so on are not as important for the humans. The route taken by the cloud does not pass through inhabited human paths but traverses the landscape aerially, over forests and mountains and at considerable speed. The cloud is also embodied, though only "a puff of smoke". It can be punctured by the sharp gems in bracelets of the divine damsels (*Mgdt*, verse *Pū.* 64) or create bristling in the *ketaka* bushes by its touch (*Mgdt*, verse *Pū.* 26). This perspective is refreshing that it is one of the reasons why this is known as one of the popular poems and has inspired similar poems of nature as messenger. A poetic work with each line immortalised within another poem was composed by the poet Jinasena to honour *Meghadūta* (Kale 1974, p. xvii).

9.2.3 Descriptions and Images of Nature in Meghadūta

Let us begin with the descriptions of the non-human beings, animals and birds. Many varieties of birds and animals are not mentioned but description of flora and fauna are regularly given in the poem. The poet using the *yakṣa*'s eagerness to give the messenger pleasant experiences along his travel, describes trees, birds, and animals engaged in different activities. For instance in *Mgdt*, verse *Pū.* 11:

> And so on hearing that thunder of yours which is grateful to the ear and which has the power to make the earth covered with mushrooms and fertile, the royal swans eager to go to the Lake *Mānasa*, and having a stock of bits of shoots of lotus shoots of the lotus-stocks to serve them as provisions on their journey will become your companions in the sky as far as the Mountain *Kailāśa* (trans. Kale 1974)

For Kālidāsa, nature is not an impersonal backdrop for his plot and human emotions of love and longing.[2] Nature is embedded in the very fabric of the narrative and poetry, influencing and reflecting the mood of not only human beings and sentient beings but also other aspects of nature such as trees, rivers, clouds, or mountains. Nature is integral to all experiences. We know that human–nature relationships are very important for Kālidāsa. His favourite heroine, *Śakuntalā* of

---

[2]As compared to his poem *Rtsm*, this poem is probably written by a poetically mature and older Kālidāsa (Kale 1974).

the famous play *Abhijñāna Śākuntalam*, nurtures the trees and plants around her hermitage and cares for them as children. In return, nature bears her wedding gifts of clothes and jewels on trees such as fruits, returning her love.

The geographical descriptions begin in verse *Pū.* 13 when the *yakṣa* addresses the cloud and says—*mārgam tat sruṇu*—listen to that route (you have to take). The route takes the cloud through the hills and plagues, forests and river paths across the river valleys of Vindhya mountains, passing beyond the Sindhu River (the Indus) on to the plains of Kurukṣetra. From here, the cloud moves northwards to the Ganga river valley and follows the river upstream until it reaches the Krouncha mountains, a pass that marks the entry to the Kailāśa-Mānasasarovara regions. The descriptions of the path the cloud till it arrives at this pass are that of real places and geographical features that are terrestrial. After the cloud enters the sacred landscape of the Kailāśa-Mānasasarovara region, the descriptions of landscapes become somewhat imaginative and fictional. For example, there are lamps that do not have a flame but are lit by lustrous gems, and wish fulfilling trees in plenty, as well as the opulence of Kubera's city, the deity being the keeper of all wealth.

The different kinds of terrain that the cloud has to traverse over are described in detail. The cloud is to begin its journey and move over a high plateau called Māla, then on to the mountain Āmrakūta (verses *Pū.* 16–17), then the cloud follows the river Revā in verse *Pū.* 19:

> Having rested on it (i.e. Āmrakūta) for a short while, (that mountain) whose arbours are enjoyed by the wives of the forest dwellers, (you will) move faster on the path, being reduced (in weight and form) due to the discharge of water (rain). You will see the Revā (river) scattered into many streams on the rugged rocks at the foot of the Vindhya mountains, looking like the (white line) drawings that decorate an elephant's body (trans. by the author).

We also see an instantiation of the aerial view of the landscape, a cloud's eye view. The foaming white streams of the river flowing against the backdrop of the grey rocky landscape are like the white decorative lines that are painted on the elephant's body.[3] A human view from the ground would see the river surface and the rocks but not the overall picture that appears like the elephant. Similarly, verse *Pū.* 18 also describes the dark cloud resting on the mountain which is white on its slopes and the picture is compared to one of breasts of the earth. Again, the view is expansive, seen from above the earth's surface, an aerial mapping in literary imagination. There are other regions described next such as the region around the rivers Vetrāvatī, Narmadā and Sindhu. A detour for the cloud would be beneficial, to visit the city of Ujjain and the temple of Mahākāla (a form of Śiva) there.

There are many references to climate and the seasons in the poem. It is found in the description of all the activities of the cloud. The arrival of the thick dark cloud signifies the arrival of the monsoons marking the end of a hot summer. There are gentle breezes that carry the cloud along (*Mgdt*, verse *Pū.* 10) the presence of the

[3]This is a common practice in India, and elephants are painted on in variegated colours during festivals and special processions.

waiting *cātaka* bird (that awaits the first shower of rainfall) all of which are associated with the season of rain. The cloud is requested to put out the mountain fires in exchange for the rest it takes on the peak. The forest fires indicate the heat of the summer and other activities such as freshly ploughed land in the Māla region that precedes the arrival of the rain (verse 17). Similar common signs of monsoon season such as the call of peacocks, the hail storms, the flowering of *ketaka* bushes and ripening mangoes are all indicated in many verses.

Intertwined in all these activities is also the season of love and passion that intensifies with the rains. Culturally, this is often described as a season of love-making, the arrival of rains is a time for love.

9.2.4 Bāna: The Ecotopia of Human–Non-human

Bāṇa is regarded as the one of the finest prose composers (*gadya-kavi*) of Sanskrit literature and he was the court poet of King Harśavardhana of Kanauj around the 7th c. A.D. (Shipley 1946, p. 459). He is well known for his prose composition *Harśacarita* and the literature text *Kādambarī*. We know of his life from the text *Harśacarita* where he recounts his own biography. Both of these works are incomplete, and *Kādambarī* was completed by his son, who reveals all the suspense and the mysteries that Bāna builds up in the first half. Bāna's clever use of words and rich descriptions is extraordinarily praised in *kathā* literature. A *kathā* is a prose composition in which a main story is told with many interrelated sub-narratives. The theme of stories within stories is popular in Indian literature (Keith 1996, p. 320). Logan (2011) writes of the masterpiece:

> Bana's other great prose masterpiece is his *Kadambari*, one of the most celebrated examples of the 'story' genre. Far more narrative-driven than the *Harshacharita*, the *Kadambari* is a sprawling romance with an intricate plot involving multiple sets of separated lovers, past births, talking parrots, apparent deaths, and miraculous resurrections (p. 33)

The plot of *Kādambarī* is derived from earlier versions of the mega story *Bṛhatkatha*, but Bāna is said to have expanded this story and added many more characters and subplots not found in the original. The story is too long and complex to be summarised here, so a shorter summary of the first part of the book that is to be discussed in this chapter with an ecocritical perspective is given below.

An extraordinary parrot is gifted to King Śūdraka, who is the ruler of Vidhiśa. The parrot is unusually gifted with intelligence and speech. It is able to express itself and is a trained scholar. The king questions it about its life and the parrot begins to narrate its own story. It gives the king a rich description of the forests around Vindhya mountains, the Lake Pampa that is found there and the old, tall tree that was its place of birth and infanthood. The parrot also tells the king of a great hunt of the Śabara tribes that led to the death of its father at the hands of an old hunter. The baby parrot which was nestling under the wings of its father was thrown to the ground and escaped alive by hiding in some leaves. The parrot whose wings were yet to develop the strength to fly, tottering on its way to the lake, was seen and rescued by a young ascetic, the son of the great sage Jābali. Then, a description of the sage's ashram is given. The sage on seeing the poor young parrot mentions that the parrot is only paying for its earlier misdeeds by undergoing this birth. The narrative voice in the story changes and the sage Jābali relates the story of the past life of the parrot to his son and disciples. This is now the main story of the heroine, *Kādambarī* and her love for a great prince *Chandrāpīda*. She has a friend Mahāśwetha whose story also gets intertwined with *Kādambarī*'s. The plot is suspense filled with the lovers being separated, and then finally, the various subplots lead to the end when all the protagonists are happily reunited with their loved ones. The parrot regains his human form as Mahāśwetha's lover, and Śūdraka turns out to be Chandrāpīda reborn.

For our purposes, the first two sections of the story make for a very significant understanding about the imagination of nature in Bāna's work. The parrot's narration of the story is significant from an ecological viewpoint. The description of the forest, the lake, the tree, and the hunt is from the vantage point of a fledgling parrot. Unlike Kālidāsa for whom nature has no voice but only action, Bāna uses the parrot to give a "voice" to nature. The parrot is also gifted with speech, intelligence, and sentiment. Whatever the reason for the parrot's extraordinary ability, the parrot as a fledgling thinks from a parrot's viewpoint. The voice of the parrot humanises it only in the sense that it displays rationality and sentiment similar to the human beings, and it still displays all the natural qualities of a bird such as eating fruits and mourning its own parrot father.

9.2.5 The Forest, the Lake, and the Tree

The description of the forest in *Kādambarī* is based on the concept of an *Āraṇya.* There are trees, creepers grasses and plants, birds, and beasts and also there are hunting tribes and ascetics who live in it. The human and the non-human, natural, and supernatural beings all inhabit the forest world of Bāna. The forest is perceived as vast and covering a large area: "The great *Vindhya* forest touches the wooded shores of the Eastern and Western Oceans and adorns *Madhyadeśa,* it is like a girdle to the earth" (Rajappa 2010, p. 19), declares Bāna through the voice of the parrot. The description of this forest mentions the activities (sometimes mythical) of cranes, elephants, parrots, peacocks, rhinos, bears, deer, boars, and even forest deities. While the forest seems to be a home to all these beings, it is yet not free of the dangers of predators. Lions and tigers create fear in all the animals as do the sounds of the Bhil and Śabara hunters. It is a place where death is on prowl: "Death lurks close, rows of tiger pugmarks, bands of wandering rhinos, wild buffaloes roaming, makes it the city of Yama, the god of death" (trans. by the author).

The forest of the Vindhya mountains is associated with a sacred geographical narrative that Bāna includes in his description. The place was known to have Sage

Agastya's ashram on the Pampa Lake. The *Rāmāyaṇa* inspires the geography of the forest. The place called *Pañchavatī* within this forest is also known to be the place where *Rāma*, *Sītā,* and *Lakṣmana* spent their time in exile. The story has entered the very being of the forest altering the fauna and the plants. It is also the place from which *Sītā* was kidnapped and *Rāma* sorrowed for her loss. These stories are woven into the description by the poet: "The deer hearing the thunder of the clouds imagine it is the twang of *Rāma*'s bow and look for him, forgetting to eat the grass half cropped, yet (unchewed) in their mouths" (trans. by the author).

Again the blood of the *rākṣasas* killed by *Rāma* is soaked up by the roots of the plants whose young leaves are blood red in colour. There is also mention of a painting of *Sītā* drawn by the sorrowing *Rāma* that the hunters find on the ground. May be it was *Sītā* herself revisiting the place because *Rāma* lived there, claims the poet. The poet creates the markers of the sacred geography through his literary imagination.

After this, the description of Pampa Lake and the vegetation, the various activities around the water highlight Bana's almost idyllic imagination of the landscape. The poet zooms in on the Vindhya forest, the Pampa Lake regions and then on to an individual *Śālmali* tree near its banks, drawing the reader into the landscape. Humans are not alien to the forest landscape in Bāna's composition. The *Śabara* women and the ascetics make use of the waters of the vast lake as do the birds and animals. The leaping fish, the peacocks and peahens, the swans (ruddy goose), and the elephants share the same place, each engaged in activities that are somewhat essentialised in literature.

There are some stereotypes of animal nature, myths, and behaviour in Sanskrit literature. For instance, elephants have pearls in their forehead and are hunted by lions. These pearls could be found stuck to the claws of lions. Swans could separate milk from water. The *Cātaka* bird feeds only on the raindrops and sits with its beak open to the sky. The *Cakora* bird loves the moon light and so on. References to these common beliefs are found not only in Bāna, but are prevalent across many compositions of those times. The literature of the period must have upheld many of these beliefs that were prevalent and these stories must have also popularised them.

The ancient silk cotton tree is described with exaggeration, very old, and very tall, "its top is lost in the clouds". A great many birds lived on the tree free of the predators that may hunt them on the ground.

> On it were countless nests, built at the edges and junctions of the branches, inside the hollows of the ancient bark and amidst the foliage, built with ease and confidence because of the availability of unlimited space (trans. Rajappa 2010, p. 26).

9.2.6 The Hunt

After the description of the daily activities of the parrots on the tree, the parrot narrates its own story. Being raised by its ageing father, its mother having died earlier, the fledgling bird is not yet old enough to fly or take care of itself. One of the most poetic passages in *Kādambarī* of a beautiful morning sets the stage for a fateful drama that is to occur in the forest. The parrot recollects its fear and the curiosity as it listens to the commotion of a great hunt as all animals in the forest panic. We hear the conversations of the hunters tracking the animals such as elephants, deer, lion, and the boars. All this is heard and not seen as the fledgling parrot is hiding under its fathers wings:

> Do you get here the fragrance of the lotus plant crushed by the charging leader of the elephant herd? Here is the smell of the *Badramasta* grass chewed by the boars.... Do you see the slushy footprints of the boars and the grass green saliva of the ruminating deer that has so recently fed on this tender grass... (trans. Rajappa 2010, p. 29–30).

The parrot can hear the hunt below the tree canopy from the sounds of elephants being killed and the panic calls of separated families, lost young ones, and scattered groups of all animals. The eerie calm after the hunt is finished adds to unnatural state of an otherwise noisy jungle.

The scene of hunting by the dark skinned forest dwellers seems to portray the poet's own stance regarding the idea of hunting. The hunters are called uncivilised, and the poet's moral position is clear in the baby bird's lament: "Alas the life of the Hunters is one of ignorance.... They destroy the very forest they live in" (trans. Rajappa 2010, p. 34).

Nature accuses the human hunters—through the voice of the helpless baby bird —those who have no compassion for the living beings or the beauty of life in the forest. We are also told here that the purpose of the hunt is not only to gather flesh, but also to kill the lions for the precious elephant pearls stuck to their claws. Does Bāna suggest here that eating meat as food is a lesser form of violence than killing animals for trade? Though indirectly, the effort of the poet to describe the destruction of the forest by these hunters seems deliberate. An idea that seems to be foregrounded is that culture and education would create human beings who would live in harmony with nature appreciating it bounties. In this description of the hunt, the poets own preference for non-violence is shown and his love for all animals in the forest can also be read into the passage.[4] The very fact that the poet takes a bird's view of the incident is noteworthy. The description of an old hunter Śabara desperately seeking flesh has an almost accusatory tone, as the parrot, now peering out from its father's wings sees a needless act of cruelty towards defenceless young and old flightless birds. In an instant, the Śabara wrings the neck of the young parrot's father and throws down the body from the tree, the young bird still in the folds of its father's wings. The young bird struggles to survival. Bāna seems to have observed the will to survive among animals and birds and chooses to express this through the young parrot's voice. Though deeply grieving for its father, it hides to save its own life and then struggles towards the lake. One might imagine the poet under a tree watching a fallen bird struggle to walk, almost tottering and struggling to safety.

---

[4]In a much later part of the story, a hunt by the prince and his army is however described as valorous, not wasteful.

> I had not grown my wings fully as yet; nor was I steady on my feet. With the result I fell again and again on my face or on my side. Supporting myself with one wing I dragged myself on the ground. Unused to such exertions I was soon worn out. At every step, I would stop, raise my head and take laboured breaths. My body became ashen with dust (trans. Rajappa 2010, p. 37).

9.2.7 The Āśram Ecotopia

The struggling bird is seen by a sage's son and rescued. It is then taken to the ashram of the great sage Jābāli. The hermitage as an ecological landscape is in between the *Nagara* (city) and the *Āranya* (the forest). There are large groves of trees of different kinds. There are also flower and fruit bearing trees and creepers. In stark contrast to the dangerous forest, the presence of nature here is benign. The difference between the two kinds of forest dwellers the ascetics and the Śabara hunters is illustrated in many different ways. The compassionate ascetics are portrayed as living in harmony with nature. The sage's son stops to rescue even a small bird and the deer which have been fed by the ashram students follow them to the lake. In Bāna's world, the deer repay the kindness of the students, "In a show of affection they would often dig up the mud with their horns for the hermits to use when they bathed. Some of the mud was still stuck to the tips of their horns" (trans. Rajappa 2010, p. 39).

Bāna's description of Jābāli's ashram is a type of cultural ecological utopia or ecotopia. We find that in the ashram, there is a picture of a community where human beings and nature live together helping each other. Like all descriptions of utopia, Bāna's imagination of a human–nonhuman inclusive is certainly idealised and fictional. The hermitage humans feed and nurture the animals. In return, the animals perform all kinds of help to the ashram residents. Parrots repeated the Vedas along with the students, friendly monkeys help old and blind hermits to find their way, elephants fetched water in their trunk to water the plants, and the peacocks fanned the flames of the sacrificial fires.

The ecotopia is not based on utilitarian values or the mutual benefit cooperation between humans and animals. The premise is that the sage's spiritual activities create a moral atmosphere where the animals and the humans are able to understand and appreciate one another. The parrot's tale recounts the sensitivity of the ashram people who treat the animals as dignified life forms and not as means to an end, unlike the hunters whose purpose is to utilise the meat and the pearls. The ethics of compassion and respect for life form a background to this discourse. Parasher-Sen (2012, p. 134) writes that there was a separateness between the forest tribes and civilised society right from the Mauryan period. Historically, Bāna who lived much later must have been influenced by his society while composing his work. It also seems that though Bāna's writing reflects the social mores of his times, his values towards nature are guided by principles of ahimsa and respect for life.

9.2.8 An Afternote

An important sociological point must be made here, particularly related to the pre-modern culture of India. This critique is both ecological and social. The times that Bāna lived and wrote were seeped in rigid social hierarchies. Forest dwellers and hunters were also stereotyped as hunters and cruel villains. The hunting of the Śabaras, as opposed to the hunting of kings and warrior classes, is often portrayed as morally reprehensible. They were included in narratives, but the various rules of social hierarchy were preserved even within the literature.[5] Thapar (2002, p. 56) too suggests this: "Initially the forest was the habitat of those regarded as outside of the social pale. Subsequently, the establishment of hermitages in the forests and the preference of ascetics for forest retreat led to some romanticizing of the forest".

Ecocriticism, a discipline that has its origins particularly in American literature, is yet to develop concepts that can address non-Western, non-English literature. When we try and understand nature within pre-modern Sanskrit literature in India, where the influence of the Romantic Movement is completely absent, we find that to refer to these compositions as nature-centric or human-centric is also difficult. A more nuanced way of understanding nature in pre-modern poetry could be through the two categories of action and natural behaviour of the non-human world. As we have seen earlier in the poetry of Kālidasa, the cloud's activities in the rainy season are described rather accurately.

The discourse of *dharma* and *karma* of nature in the literature is one of the ways in which we could analyse nature that is inclusive of the human being. In the literature, the non-human is included in the human world. Nature is seldom passive in these narratives; nature participates in the human world by being itself, active as nature. Winds blow, clouds rain, the forest fires burn, and mountains stand tall. All these are *dharma*, the appointed duties of natural things or the order of nature. *Karma*, as actions of nature objects, is completely instinctive and in sync with their *dharma*. It is through these categories that the feminisation of nature or the silencing of nature takes place.

The descriptions of human engagement with nature are also based on *karma* and *dharma*. The adaptation to seasons and landscapes, the ethical and moral attitude to the non-human part of the world are all based on activities and the order of nature. The literature discussed in this chapter also in many ways reflects these important conceptualisations from Indian philosophy.

---

[5]The social stereotype image of the "tribal" is often reflected in current times where wilderness activists see these people as trespassing on "nature" and often accuse them of poaching animals. However, the soft exploitation of these areas by so-called legitimate activities such as tourism, building roads, dams and so on is considered to be cultured and permissible.

References

Abram, D. (1997). *The spell of the sensuous: Perception and language in a more-than-human world*. New York: Random House publications.

Berriedale, A. B. (1996). *A history of Sanskrit literature*. New Delhi: Motilal Banarsidass Publishers.

Devadhar, C. R. (2001). *Ṛtusamhāra of Kālidāsa*. New Delhi: Motilal Banarasidass Publishers.

Kale, M. R. (trans.). (1974). *The Meghadūta of Kālidāsa: Text with Sanskrit commentary of Mallinātha, English translation, introduction, notes, appendices and a map*. New Delhi: Motilal Banarasidass Publishers.

Logan, P. M. (Ed.). (2011). *The Encyclopaedia of the novel* (Vol. 1). Malden, MA: Wiley.

Macdonell, A. A. (1927). *India's past: A survey of her literatures, religions, languages and antiquities*. New Delhi: Asian Educational services.

Parasher-Sen, A. (2012). Of tribes, hunters, and barbarians: Forest dwellers in Mauryan period. In M. Rangarajan & K. Sivaramakrishnan (Eds.), *India's Environmental history—A reader* (Vol. 1: From ancient times to the colonial period). New Delhi: Permanent Black.

Rajappa, P. (2010). *Kadambari:Bana*. New Delhi: Penguin Books.

Shipley, J. T. (Ed.). (1946). *Encyclopedia of literature*. New York: Philosophical Library.

Tandon, R. (trans.). (2008). *Kalidasa Ritusamharam: The garland of seasons*. New Delhi: Rupa & co.

Thapar, R. (2002). *The Penguin history of Early India*: From the origins to AD 1300. New Delhi: Penguin Books.

Chapter 10
Relating to Nature: Worship, Care, and Ecological Ethics

Abstract This chapter describes some of the possible ways human beings can perceive, based on alternative conceptualisations, narratives, and imaginations of nature described earlier in the book. The human being within these world views is not seen as separated from nature, but is seen as an embodied, en-worlded being who is connected to other beings and the world in an ethical relationship. These relationships are understood through moral actions towards the environment given by some central principles of ethical thought such as righteous duties (*dharma*) and non-injury (*ahiṃsā*). This chapter concludes with a section on the idea of conservation and ethics with a pragmatic note on the conservation practices of the Bishnoi community.

Keywords Relating to nature · Ecological ethics · *Dharma* · *Ahiṃsā* · *Non-violence* · Lake privatisation · Theory of action · Disinterested action · Conservation · Trusteeship · Bishnois

10.1 Introduction

As seen in the earlier chapters, it is possible to describe conservation as "change," rather than as non-destruction or preservation. To lend clarity to the ideas that I will discuss in this chapter and the next, I will rely on my experiences of a cultural geography project I worked on, using a real-world illustration of a "lake restoration" project in an urban area.[1] Recently, the Lake Development Authority (LDA), Bangalore, came up with a plan for lake development. Under this scheme, lakes would be leased out to either a private contractor for 15 years for a hefty annual fee or citizen groups (resident welfare associations, educational institutions, etc.,) for 5 years. They claimed that leasing out a few of these lakes would pay for the

[1]An urban studies' project on 'Lake as Urban place' was carried out by the researcher along with two of the colleagues through a stipend grant from SARAI, Centre for the Study of Developing Societies, New Delhi. See Baindur (2014).

M. Baindur, *Nature in Indian Philosophy and Cultural Traditions*,
Sophia Studies in Cross-cultural Philosophy of Traditions and Cultures 12,
DOI 10.1007/978-81-322-2358-0_10

maintenance and development of other lakes under the control of civic bodies. The private bidder who is granted the memorandum of understanding (MoU) would "develop-operate-transfer" (DOT scheme) the lake after using it for 15 years. The Hebbal Lake in Bangalore, Karnataka, is one of the lakes that had recently been handed over to a private company under this scheme. Another lake close by, the Nagavara Lake, has already been "developed" as an entertainment area by a private company. This is one of the typical scenarios one might face as an environmental ethics problem.

Conservation of this water body has been interpreted differently by the government-nominated authorities, the residents of a local area who had taken care of it in the past, by users of the lake such as park-goers and contract fishermen, conservationists who supported the cause of aquatic birds, and the private company that had bagged the contract for lake development. From this description, it is clear that even a small conservation effort, such as that of a lake, is not a simple issue to be resolved within the current paradigms of environmental ethics. As I work through the conceptual foundations of this chapter, I will share insights from this study of the lake conservation/development project to explain and illustrate how conceptual perspectives create different frameworks of moral action, and some of those perspectives by virtue of being richer and more varied help open up new areas of understanding the problem itself.

In Indian thought, the relationship between the being and its environment can be understood through three categories of phenomenal experience called *ādhyātmika*, *ādibhūtika*, and *ādidaivika* interpretations. Under this schema of understanding, as embodied selves (*ādhyātmam*), the relatedness is given by encounters between the various faculties of sensing and action *(indriya*s), in combination with the mental faculties and the objects of experience in the phenomenal world.[2] Both disinterested action and restraint are moral in nature at the level of an embodied encounter of human beings with the world. These have significance for the ultimate goal (*mokṣa*) and the proximate goals (*dharma, artha,* and *kāma*) of human beings. The second level of encounter (*ādibhūtam*) occurs between the embodied human being and other beings. In this type of encounter, the relationship of the beings is given by the framework of predispositions—the *bhāva*s—and the predominant *guṇa*s. In understanding the world as "*nisarga*"—as levels of created beings and functions—human beings are placed "within" the realms of creation, allowing for eco-moral responsibility in the form of *karma* as moral action.

The third level of encounter (*ādidaivam*) is between the human being as part of the larger cosmic order which she belongs to. Here, the relatedness is construed as the relationship of *dharma and ṛta* expressed as moral action. A phenomenal relatedness is created by the construing of human beings as caregivers to the earth as an ageing mother and as "beings that shelter" the various beings (including the environment-beings) through a larger interpretation of *ahiṃsā*.

---

[2]Sāṁkhya philosophy calls these various faculties as thirteen instruments. See Gerald James Larson, *Classical Sāṁkhya: An Interpretation of its History and Meaning*, 190–191.

The lake, which we can regard as a representation of "nature" in the everyday experiences of the people who use it, is not a blue water patch on the ground, nor is it a space to be filled and developed. For the multiple users of this lake, the concrete materiality of the lake—the banks, water, fish, birds, and the weeds—is tangibly encountered as *places*. Encountering the lake on the ground is a plethora of *place experiences*. Each group of users acts differently towards the lake given the relationships of the people to the lake, and each other is different in each case. Why would that be? I suggest that their ethical stance is given by the conceptualisation of the lake or put in a broader perspective, the conceptualisation of nature itself.[3] Here, on one level, the users of the lake are concerned with the various benefits they derive from the natural resources, yet on another level, they worship the small goddess shrine, which is dedicated to a water goddess. The lake is thus used for religious purposes during the festivals. As mentioned in earlier chapters, the divine coexists with the material, but the deity, though connected to lake, seems to be dematerialised from the lake itself. While all the beliefs and narratives of the lake users were clear to some extent from talking to them, their actions towards the lake and what they valued about it differed. Are the everyday actions of all these users of moral importance? If a visitor just looks at the lake and enjoys its view, is she performing a moral act? Or is it only the action of the private company that drained the whole lake in two days evaluated as a moral act that is "wrong"? What about the dhobi who washes clothes in the lake? What is the ethical evaluation of the boating activity? From an environmental perspective, washing clothes "pollutes" the lake as much as the boating does. Authorities however claimed that the money collected for the boat ride may "compensate" for the activity. They also claim fencing the lake and charging admission will help "conserve" the lake from degradation. On the other hand, due to all this development and fencing, all other direct users of the lake—the fisherman, the dhobi, or the cattle feed collectors—will lose their source of livelihood. What actions are environmentally correct? What can we say about the concept of an "environmentally ethical action"? The lake is used as a source of water by the local users, for washing clothes or cattle. The embankments that encircle the lake have rich marshy grass that is cut for feeding cattle. For yet other urban residents who live around this area, it is a picnic spot and the water merely a backdrop for recreational boating. Each user of the lake is "related" to the lake from all the three levels. While some have "embodied" experience of the vision of the calm waters, the lake itself is a direct livelihood support for some of these users. One could posit from these various categories of relationship that people can and do relate to nature and natural objects through a utilitarian perspective ranging from direct use of the tangible material of the lake to indirect use of the lake as an aesthetic backdrop.

Given the very different conceptualisations of nature in Indian thought, it is possible to posit a theory of action that provides a very valuable ethical framework for the praxis of human behaviour towards nature. Light and Rolston III (2006) note that one of the proper conclusions for an environmental ethic is the grounding for

[3]For a detailed report of this study, see Baindur (2014).

action: "A legitimate goal of ethics is to provide us with a language, with effective arguments, whereby we can claim some kinds of action are right or wrong, or at least better or worse, independent of their cultural and legal context" (p. 6).

In the argument for the creation of an "ethical language", I find that the requirement to have concepts that are "independent" becomes a problem, especially within Indian thought where culture and philosophies are strongly interwoven with religious and social realities. Before I propose my ideas on a theory of action in relationship to ecological ethics, I would first like to summarise Larson's (1987) critique of deriving conceptual resources from Asian traditions of pre-modern thought for current problems and address some of these issues during the ensuing discussions.

Larson (1987) critiques the project of deriving conceptual resources from ancient philosophies to solve current environmental problems by describing three considerations that are fallacies. First of these, according to Larson, is the idea that there is an asymmetrical relationship between the conceptual ideas of the past and the ecological crisis of the present. Using the metaphor of milk and curds, he argues that the change and evolution cannot be undone:

> To use a *Sāṁkhya* idiom from South Asia, we have reached the "curds" stage of an earlier "milk" stage in the unfolding of our natural habitat as a species, and we cannot wish away the curds stage by arguing for a symmetrical relationship between milk and curds (Larson 1987, p. 155).

The second fallacy that Larson (1987) points out is that of "disembedded ideas". He points out that "conceptual clusters" can occur only within frames and paradigms that are based on cultural or historical backgrounds. To extract a concept from a particular framework of ideas, and use it elsewhere independently without any relation to its origins, is a fallacy. He argues against the notion that concepts "can be disembedded, dusted off, and somehow utilized in dealing with an environmental crisis" (p. 156). Concepts, he insists, must be studied as they function within their frames and paradigms culturally and temporally.

The third fallacy Larson (1987) refers to is "the fallacy of the sovereignty of the subject." He points out that our interpretation of the methods of self-awareness and the deeper understanding of who we are as human beings may be outdated. Taking a cue from Foucault, he writes:

> … Future progress in dealing with the environmental crisis will have to address critically the problem of 'the sovereignty of the subject' and to avoid the fallacy of assuming that what we think we are is in any sense an adequate, consistent or a clear measure of what we are (p. 157).

As we have seen earlier in Chap. 8, an important aspect of the perception of reality in Indian is through the concepts of *ṛta* and *dharma*. One of the difficulties around the ethics of action here is that questions like right and wrong or evil and good do not seem to attract critical debates in Indian philosophy. This is because within the divine ordering of the entire cosmos, actions appropriate to one's place order (by one's *dharma*) are already prescribed. Bilimoria (2007, p. 34) compares this to the conception of natural law in the Western tradition and states that the absoluteness of the moral law called *ṛta* ensures harmony in the universe. We have seen in Chap. 8 how among the four ends of human life, the *dharma* that is determined by one's form, role, function, and innate nature is common to all created beings and even the world of materials. In this sense, *dharma* is almost like a natural law. Each created being is almost impelled to act as per its *dharma* in order to maintain the stability in creation. The *dharma* of rivers for instance is to flow downwards and that of fire to burn. *Dharma* ensures members of groups and community follow and act appropriate to their group's position and status in society. While the classes of other beings have the most basic of the *dharma*s given by their instinct and innate nature, human *dharma*s are much more complicated and bound to moral choice and action that in turn depend on one's place in the socio-religious hierarchy.

The category of moral or ethical choice is thus limited to performance or non-performance of an ordained action. Or to use an Indian phrase, it is about a *dharma* (righteous) or an *adharma* (non-righteous) action or inaction. This *dharma* is a given, dependent often on scriptural or traditional prescriptions. They advise simply, "do this" or "do not do this". Following these instructions is being moral, and not following them amoral. In this framework, all we are left with is the action itself, not the choices. For instance, if one is a soldier, one's *dharma* is to fight and kill the enemy. One's *dharma* is given by one's "being in the world" so the choice of "not-killing" implies not "being a soldier". So, if a soldier runs away, he has failed his *dharma*. Enmeshed in the notion of *dharma* is also the action of fulfilling one's words or promises.[4] By one's being in a particular position within this socio-cosmic order, one's word to fulfil one's role in the cosmos is already given, a promise made by origin, so to speak. To fulfil *dharma* is a non-verbalised promise. This can change only if one's being changes, altering one's karmic path.[5] The moral choice of "not-to-kill" can be only made, if the soldier decides to change his socio-cosmic position and the teleological purpose of his life—by becoming a *yogi* through renunciation or by the deliberate loss of his social position (being outcaste). We see here how moral choice in its popular conceptualisation is difficult to carry over from the Indian tradition. So in the following section, we shall examine the idea of *karma* and moral choice in detail.

10.2 Action in Indian Thought

Morality based on metaphysical grounding in philosophy and religious mythology and an explication of ethical norms through a theory of action in Indian thought can be distinguished. Chapple (2007), for instance, differentiates these two trajectories clearly: "Teachings on action establish ethical norms; stories of rebirth offer cosmological explanation" (p. 355).

---

[4]Entire mythologies are based on the idea of keeping one's word—King Harischandra's story is well known.

[5]Taking the path of renunciation is one such possible change; sometimes, divine beings are also cursed to lose their position.

He also recognises the unity about the idea of *karma*, or moral action, underlying the liberation theory of many schools of Indian thought such as Sāṁkhya, Vedānta, Jaina, or Buddhist:

> Yet despite this ideological disharmony (well-recognized and widely debated amongst respective schools), on the level of praxis, there is an apparent agreement. All the schools of Indian thought emphasize moral action as an integral part to liberation (pp. 353–354).

The problem that most environmental philosophers are likely to confront in Indian philosophy is around the idea of liberation that seems to guide all moral and religious action. Since the notion seems so culture-specific and world denying (as in the case of Vedānta for instance), the end goal of liberation for them seems to be unsuitable for environmental ethics. How can such an ethical framework inform a general cross-cultural or even a modern world view in India? How can cultures and communities that have no belief in the idea of liberation understand moral action? If they still accept the concepts around moral action theory, it would still fall under the fallacy of disembeddedness that was raised by Larson.

Therefore, I first discuss how the idea of moral action in Indian thought need not be perceived as embedded in mere soteriological purposes of a human life. Though the ultimate purpose of human life is *mokṣa*, or liberation, the path to liberation is through proximate moral action in the real world.[6] It is also true for these traditions that moral action is not based on an ethics of moral consideration of the recipient of the moral action, as much as it is based on the positive or negative impact (whatever that may be) it has on the moral agent.[7] We can therefore say that the proximate purposes, which guide everyday action, are bound by rules of *dharma*, while all these actions ultimately lead to *mokṣa* in this lifetime or other lifetimes to come.[8]

What then is a general theory of action that is conceptually free of an ultimate goal of liberation? While most Indian philosophies differ on finer points of metaphysical and epistemological frameworks, but in a general way, all of them deal with the idea of the restraint of "desire" and attachment to sensual pleasures as an important norm for obtaining both the proximate and the ultimate goals of human purpose. The ultimate and proximate purposes are not set in opposition, but can coexist in the same space.

A theory of *karmavāda* or action forms the basis of a number of discussions on moral philosophy in Indian philosophical schools. This is because the field of ethics in Indian thought seems to be largely normative and connected with action. Mohanty (2007) states:

---

[6]The definition of liberation differs in all the different schools of Indian thought and would be too long a digression here. For a summary, see Dasgupta (1922, p. 4).

[7]Moral considerability for acts of virtue like charity or kindness is also designated as '*pātratva*' or deservingness.

[8]Actions do not directly lead to *mokṣa*, but lead to mental purification leading to knowledge. See Dasgupta (1922, p. 76). It is also important to remind readers here that most traditions of Indian thought subscribe to the doctrine of transmigration (rebirth).

> Legal thought and ritualistic speculations, ethical and spiritual philosophies centre on theory of action. Even those spiritual philosophies which recommend inaction or transcendence of action in any of a whole variety of senses make use, in doing so, of certain widely held ideas about what action is all about (p. 57).

However, at the individual embodied level, each act is seen as a complex causal process of cognition, desire, willingness, embodied expression, and performance. Mohanty (2000) unpacks this scheme of action and posits a general framework for all action in Indian tradition. According to this framework, "Knowledge (*jñāna*) produces desire (*cikīrṣā*), desire produces the will-to-do (*pravṛtti*), this in turn produces the actual motor effort (*ceṣṭā*), the consequence of which is the action (*kārya*) as a completed performance". Mohanty also adds that the concept of an agent (*kartā*) is important in this schematic analysis.

What is interesting about this schematic analysis of embodied, en-worlded action is not the description of "how action occurs", but the fact that "action can be restrained" at any of these levels of the causal chain or that its direction/implementation can be changed at any point. Since all of the action is guided by moral order and law, there is no action that is not within the evaluative framework. Another important part of action theory is *karmaphala* or the idea of the result/fruit of an action. The causal chain for human or divine realms is different due to the cognitive component of action, *jñāna*. This means the actions of beings such as animals or plants are *svadharma prerita*, inspired by their own *dharma* (moral-natural law). It is not possible for these animals or plants "not-to-follow" dharma or cause a change in their actions through will. Though still under the moral realm, these actions of animals or plants do not yield the effects of *karma* as it does for the human beings.[9]

When we look at our everyday actions, we can clearly see the schematic framework. Now if we add the ecological viewpoint to this scheme, the question we should be asking is not if there are environmental values in moral action, but whether it is possible to have any action in the material world that is non-environmental value laden. In other words, all actions that we perform as embodied beings in the material world have a moral value, ecologically speaking.

One of the ways to look at this is through the "pleasure–good" binary of actions in Indian thought. If we shift our conceptual attention from the question "how do we exploit nature?" to "why do we exploit nature?" the proximate answer would be that within the current metaphysical grounding, we think nature is separated and inferior in value to us as human beings and therefore exploitable. However, I find that from a framework of Indian thought, the answer would be, we exploit nature for our enjoyment (actually, both desire, and enjoyment—*kāma* and *bhoga*). It is not just the idea of substantive oneness or creative play that can give us a firm foundation for moral action that impacts the environment, but focusing on the normative actions of the human being that demonstrate our relationship to nature.

---

[9]Some scriptures explain that the lower wombs are for expending the demerits and the higher realms for expending the merits, and it is only the human realm that can earn merits and demerits simultaneously to move either upwards or downwards.

The biggest norm is the reduction of desire that is recommended for both the ultimate goal—*mokṣa* (liberation)—and proximate goals such as *dharma* (socio-cosmic obligations), *artha* (material self-enhancement), *and kāma* (desire fulfilling) in the course of a human life. The moral choice is between what is called "good" or "beneficial"—*śreyas* or *hita*—path to lead a life or the pleasurable—*preyas* or *priya* —path. While both actions are not "evil", the long-term outcome of choosing the "good" over the "pleasurable" is that one gains merits and these can lead to rebirths in higher realms, spiritual development, and knowledge of liberation. We can say that action we perform in today's world is ecologically relevant as a "*priya*" or pleasurable action that exploits nature, or as "*hita*" or good action that is sensitive to our interaction with it.

10.3 Action as Desire-less in Advaita and the *Bhagavadgītā*

One of the central issues in environmental philosophy has been the perceived ill-effects of dualistic modes of thinking. Scholars categorise these as being historically derived through Western patterns of thinking. Nelson (2000) points out that one of these modes of duality—that of matter and spirit—lends itself to the idea of the elevation of spirit above matter, which further "leads to an alienation from the natural world" (p. 61).

The substantive oneness of all beings, which is one of the fundamental ideas in Advaita (non-dual) philosophy and its relevance for environmental philosophy, has been much debated in current literature. It is interesting to note that though the philosophy itself is pre-modern, it is understood and propagated by numerous religious and spiritual groups in India today. Though not all of these current practitioners of Advaita may show any deep understanding of the philosophical sub-limities of this doctrine, it would not be wrong to say that among most common people, this philosophy almost holds a hegemonic sway in India.

Advaita, as a philosophy, holds a certain appeal for the current environmental thinkers because it makes "non-dualism" its central doctrine. Balasubramanium (2000, p. 108) clarifies that the concept of non-duality creates unity between matter and spirit and provides a foundation for eco-philosophy. Advaita philosophy is based on the basic tenets of Upanishadic literature and to some extent on interpretations of the *Bhagavadgītā*. The philosophy is said to be derived from the ideal monism of Gaudāpāda, as set forth in his work called *Māṇḍūkyakārikā*. Śaṅkara advocated a philosophy called Advaita (non-dualism), deriving his central concepts from this work of Gaudāpāda. Balasubramanium (2000) notes that in a pre-Śaṇkara version of Advaita, spirit and matter are related through identity in difference (*bhedābheda*). He notes:

> Similarly, though spirit and matter are distinguished, they can never be separated such that we can speak of spirit in matter or spirit as matter governed by the relation of identity in

difference. If so the entire physical universe being a manifestation or differentiation of the spiritual Brahman cannot be anything but spiritual (p. 108).

He then clarifies how, in Śaṅkara's Advaita, *Brahman* (the supreme) is not modified into the world, but without undergoing modification, it forms the substratum or the ground or the support for the latter (p. 109). In brief, the central idea of Advaita philosophy holds that the true reality is the *nirguṇa* (unqualified) *Brahman* (the supreme) and the *jagat* (phenomenal world) is but a mere appearance of *Brahman*. Those in support of this philosophy for environmental ethics maintain that the spiritual oneness of nature as *Brahman* is very important for a metaphysical grounding for the human attitude to nature. Deutsch (1989), for instance, writes that strictly rational and empirical perspectives have made nature factual and not value based. He suggests "Nature can become value laden, it seems to me, only from a spiritual perspective which sees nature either as a manifestation of a spiritual being or, Advaitic-wise, as an appearance grounded in a spiritual reality" (p. 264). Elsewhere, he maintains that the recognition of all life as one and that everything is a reality in essence forms "the fundamental ground of all existence" (1970, p. 4).

On the other hand, this idea of the world as "mere appearance" immediately creates a deep problem for environmental philosophers, who would prefer a more non-idealistic perception to be conceptually beneficial for any kind of ecological care ethics. Kirkpatrick (2001) suggests that this concept of "world as appearance" may work against the ethics of ecological care.

Others choose to point out that the connection between the inner self and the outer world forms an important fundamental concept for Advaita. Marshall (1992) suggests that this philosophy is not completely world-denying: "While *Brahman* and the *Ātman* are one, the world is not a pure illusion. External objects are not just forms of personal consciousness. The world may be a manifestation of *Brahman*, but it still exists in a fashion" (p. 31).

Obviously, Marshall here is referring to the idea of what is called as *vyavahārika satya* or the concept of empirical reality of the world and individual soul in order to fulfil practical purposes, which is used to explain the norms of everyday behaviour in many texts of Advaita. So, in a way, despite the fact that the non-dual philosopher denies the absolute reality of the world, she does not deny the day-to-day interaction that she has with it. These practical actions however should ideally aim at gaining knowledge for the purposes of liberation and dispelling ignorance.

We find that in the current literature, however, the argument against the suitability of Advaita as a conceptual foundation for ecological thought revolves around this very world-denying *māyā* principle. Nelson (2000) argues that as Advaita is founded on the principle of transcendence that is based on liberation from forms and an attitude of alienation from the natural world is a possibility: "Advaita achieves its brand of non-duality not inclusively but exclusively at great cost: the world of nature finally cast out of the absolute, out of existence" (p. 79).

For Elliott Deutsch (1989), however, the very idea of everyday actions aimed towards liberation creates a ground of disinterested action or *karma yoga*—action performed in a state of freedom. He argues that the phenomenal interaction of a *karma yogi*—a disinterested actor—creates the freedom to engage in creative play (*līlā*). He notes:

> ... freedom, which I believe is required for natural reverence, is in the latter view not to be conceived as the opposite of necessity (from which it would then never escape), but as obtaining on an entirely different level of being—and, in relation to nature, the source of creative making which is a kind of a play (*līlā*) (p. 264).

Freedom, or *mokṣa*, is the only solid ground for that disinterestedness. This, he claims, is similar to the ground for a Kantian disinterestedness of the aesthetic egoless freedom.

The value for nature in creative play is produced in a paradoxical way from radical valuelessness in relation to absolute reality. This grounding, Deutsch (1989) claims, results in a basic attitude of reverence to nature. Larson (1987) admits that Deutsch's idea establishes a new approach to nature reverence (p. 151). Within the truth of a relatively existent world, called *vyavahārika satya*, everyday moral action is encouraged, particularly through a much popular practice of disinterested action called *karma yoga*.

Though disinterested action as an ethical norm is common to many philosophical streams of thought, it finds its full expression in the action theory of the *Bhagavadgītā* (henceforth referred to as *BG*), as explained in the Advaita philosophy. Mohanty (2000), in his analysis of the Indian theory of action, suggests that under the broad category of *dharma*, both the theory of virtues (ought to be) and the theory of duties (ought to do) are included (p. 112). Writing about the relationship between values and action, Mohanty insists that a moral theory must focus on the theory of action:

> It may, however, be that what we have is a theory of value in a wider sense, which includes moral theory inasmuch as moral values that are a subset of all values. In a specific sense, all 'ought' pertains to actions and values, which can be striven after through actions (2000, p. 116–117).

The *BG* describes that disinterested action, or *karma yoga*, is that action which is performed without *rāga* (attachment), especially to its fruits or benefits. A naïve understanding of this concept would lead us to believe that the *karmayogin*, in being driven by his transcended nature, would do any action without accountability, in a world-denying manner. The commentary by Śaṇkara clarifies the phrase in Chap. 3, verse 9 of the *BG*: "*Mukta saṅga samāchara*" (Without attachment, act!).

> The BG explains what action done with attachment is like in contrast. In verse 16: *"adhāyurindriyāramo mogham partha sa jīvathi"* [He who is engaged in doing action that merely seeks the pleasures of the senses, is living a wasted (delusional) life] (trans. author).

This brings us back to the *kāma-bhoga* (desire-enjoyment) argument that I was making in an earlier section on action. The purpose of such day-to-day moral action, in contrast to action performed just to please the senses, is that the former, according to Śaṇkara, leads to knowledge and detachment through a process of mental purification, while the latter to the endless cycle of births and deaths. The text further clarifies that the right choice is to perform those actions that promote the "welfare of the world" (*loka-saṁgrahamevāpi*) (Verse 20, Chap. 3). Following the textual reading of the *BG*, I summarise the arguments thus:

Firstly, given my earlier conclusion that all action is moral in nature, one can posit that the morality is dependent on disinterestedness in enjoyment. This is due to the emphasis that action performed for the fulfilment or attachment to a particular outcome is immoral. In contrast, the action that is performed for the welfare of the world is moral.

Secondly, moral action ultimately leads to wisdom that gives liberation; action itself is not directed to liberation. This sets the moral action free from the forced consequence of liberation, at least for proximate purposes.

Thirdly, liberated person performs action with disinterest, which is action that does not involve indulgence in sense pleasures.

Fourthly, restraint of the above kind of desire is an "ought to do" in Advaita philosophy that leads to the "ought to be" state of the *stithaprajña*.

The same idea is also mentioned in *Kaṭha Upaniṣad*. The text emphasises the idea of preferable (*śreya*) versus pleasurable (*preya*). Though human nature seeks pleasure, it is ethically important to move towards the preferable. Verse 1, canto II reads:

> The preferable is different indeed, and so indeed, the pleasurable is different. These two serving divergent purposes bind men. Good befalls him who accepts the preferable among these two. He who selects the pleasurable gets alienated from the purpose. (trans. Gambhirananda 1957).

10.4 The Idea of Conservation

Preservation can be described as the act of keeping safe from destruction, decay, or damage. The word conservation also refers to the act of preserving something, guarding or protecting, or keeping a thing in its entirety. As far as English is concerned, the words are interchangeable and can be used for each other. Two ideas coconstitute conservation as a process. The first idea is that conservation occurs in time. Anything that is conserved, say for example, a painting or a building are all conserved over a period of time. The second is the idea of preventing annihilation. In some ways, the natural state of things is to degenerate and to change over time. Things grow, decay, die, or simply turn to dust over time. Conservation in terms of a lay understanding stops these changes and halts time as it were, for the object being conserved. Moving beyond this naïve understanding of conservation, if we examine the concept itself, we find that there are certain prerequisites for conservation, especially from a scientific viewpoint.

Conservation is observed through the "before and after" of a time period. The before and after state of the conservation is compared using some property. In an earlier work with Dr. Sarukkai (2010), I have summarised the general idea of conservation: thus, the first prerequisite being the relationship of conservation to properties, and the second being the quantification of these properties:

> So in any usage of 'conservation', there is an assumption that conservation is about properties and not about the metaphysical substance; and two, that these properties are of the system as a whole—meaning that conservation is a measure of the before and after, and thus is something which 'measures' the 'before' and then the 'after' and finds these two to be the same (Baindur and Sarukkai 2010, p. 11).

The suggestion is also that there are prerequisites for these conserved quantities for science: "First of all, for science, properties are conserved and these properties have to be quantified .... Secondly, conservation of these properties is not true in all cases. This holds good only in what are called closed systems" (p. 11).

The concept of conservation when transferred to the idea of ecological nature becomes problematic on both these counts. What are the so-called properties of nature that can be conserved? Or even if we locate the properties of "the ecological nature", how do we treat it as a closed system?

The paper points out any significant notion of conservation has to engage with two problems—that of conservation of "wholes" and the conservation of "space". Both these are linked ideas of what and how much is to be conserved: "Some conceptual problems that are fundamental to conservation have to do with the idea of measuring spaces, communities, and/or populations. The primary object of conservation for some conservationists is based on the ecosystem approach that believes in saving the living spaces of organism" (p. 11). For others, it is about conserving wholes like species, populations, or biotic communities. "The conceptualisation of space cannot be reduced to areas and volumes, any more than the quality of existence of organisms can be reduced to the number of creatures" (p. 11). The main argument here is that "Conservation within ecology is linked to the idea of a conservable property of nature, even though nature is infinite and is an open system".

From these arguments, it is easy to see why the concept of conservation finds itself being redefined repeatedly. Besides, many of these "properties of nature" are in fact value-laden cultural properties. Take for instance the concept of an endangered species, which is often used for the purpose of conservation. Smith (1976, p. 34) traces the concept of "endangered" to the idea of rarity. He suggests that though ecologists see rarity as a biological concept, it is nevertheless a value-laden cultural "property" of organisms in conservation. This is because natural rarity is seldom a necessary and sufficient condition for conservation. He provides the argument that rarity is a cultural concept: "That something is rare does not imply that it must be preserved. The characteristics that distinguish as rare must also be valued. Arguments in favour of preserving an object can be based on the fact that it is a luxury, a necessity, or an asset" (p. 34). We see that it is the human perception of a particular kind of rarity, that is the basis for the concepts and buzzwords of conservation, such as endangered species, fragmented population, or biodiversity, thereby making conservation a human-centred issue.

Conservation can also be categorised from the perspective of "Conservation Biology" that is seen as a prescriptive science. Soulé (1985, p. 727) defines this discipline as "… a new stage in the application of science to conservation problems, addresses the biology of species, communities, and ecosystems that are perturbed, either directly or indirectly by human activities or other agents". He adds that preservation of biological diversity is a goal of conservation biology.

Houtan (2005, p. 1371) claims in his paper that conservation needs to move beyond the realms of a mere positivistic science that claims to be value free. He writes "… conservation needs a cultural legitimacy that inspires enthusiasm allegiance and personal sacrifices"—in other words, it needs to induce actual change in human behaviour.

10.5 Inclusive Ethics and Conservation

When one talks of conserving nature today, the usual reference is to the idea of conserving an ecological landscape, an area that is demarcated as a "nature zone" called by various terms such as national park, wildlife sanctuary or preserve, biosphere and so on. The activity of nature conservation involves not only presuppositions of the concept of nature but also questions about the nature of wilderness as construed as exclusive of human beings and its appropriateness for indigenous communities that have lived in such areas for a long time. Displacement of such communities and the intervention of the state through policy and laws have created a model of conservation that alienates the people from their environment politically and ethically.

There are two dominant viewpoints present in the history of conservation. When enquiring into what the conceptualisation of nature is, as we have seen the earlier chapters, there is the attempt to place the human either as being "in nature" or "with nature". There is always a tension between these two viewpoints and that has an impact on the way that conservation has been practiced.

Callicott et al. (1999) note the fact that there are two dominant schools of conservation philosophy and that the primary difference between them is that one considers humans as a part of nature and the other sees them as separate from nature. In the USA, for instance, these two views led to two different movements. "Aesthetic-transcendental Conservation Ethic" was propagated by John Muir, Thoreau, and Ralph Waldo Emerson. This discourse emphasises the importance of wilderness. For these thinkers, nature was sacrosanct and needed to be preserved. The human had to be kept outside, being allowed in only for restorative purposes. Wilderness could not be the "habitat" of humans.

On the other hand, people such as Gifford Pinchot and Theodore Roosevelt led the "resource-conservation ethic movement" that focused on multiple uses of nature. Logging, watershed protection, recreation, or game hunting could be encouraged in a sustainable manner with abundance of natural resources. From these two movements, we have inherited the concepts of preservation and sustainability today.

Aldo Leopold with his "evolutionary-ecological land ethic" understanding tried to bring these two streams together and provide a philosophical foundation for conservation biology. Leopold's (2006) *Sand County Almanac*, a collection of essays, posits what is often called "holistic environmental ethics". One of the essays from the book that represents this view is called "The Land Ethic". Leopold's "The Land Ethic" may be summed up in the statement from his essay: "A thing is right when it tends to preserve the integrity, stability, and beauty of the biotic community. It is wrong when it tends otherwise" (p. 24). This statement reflects the core principle of Leopold that seeks to include land within the reach of human ethics.

Leopold proposes that for conservation, a radical change is required in the human relationship to land and there is a need for an ethics that extends to the land. The extension of ethics is an ecological process. Ethics defined philosophically, according to him, is a distinction between acceptable social behaviour and unacceptable social behaviour. Ecologically, however, ethics is "... the limitation on the freedom of action in the struggle for existence". The central problem for Leopold (2006) is that the relationship of human to the land is largely economic: "There is as yet no ethic dealing with man's relation to land and to the animals and plants which grow upon it. Land, like Odysseus' slave-girls, is still property. The land relation is still strictly economic, entailing privileges but no obligations (Leopold 2006, p. 38)". The conservation movement, thus, is an attempt to consolidate the idea that humans do have an obligation towards land and its occupants (also quoted in Baindur and Sarukkai 2010).

Though biologically competitive, human beings cooperate within a community, resulting in the development of ethics. The land ethics simply extends the human community to include soils, waters, plants, or what Leopold collectively called "land". Though there is conquest and use of the land, land ethics will alter the way land is viewed, conquered, and used. Conservation would then be a state of harmony "between man and land". Leopold remarks that education alone would not alter the slow progress in the field of conservation, but there needs to be a change in the foundations of conduct in the "loyalties, affection, and conviction". Since land use ethics are governed by economic self-interest, conservation programmes do not tackle the obligation that a farmer has to his land community beyond just using the resources.

Madhusudan and Shankar Raman (2003, p. 51) also refer to these two current conservation paradigms in environmental history: firstly, sustainable development, the objectives of which were to maintain ecological services, and sustainable use of biological resources and secondly, the conservation of biological diversity. They explain that this is in contrast to the preservationist paradigm that had evolved from "the biological premises of evolutionary and functional uniqueness, and the ethical value of species, regardless of the utilitarian value to humans".

Callicott et al. (1999, pp. 23–24) suggest that unlike the earlier streams of preservationism and resourcism that seemed to be in direct opposition, the new conceptual categorisations of conservationists as compositionalists and functionalists is a complementary view. The compositionalist view is that culture separates humans from nature, and nature has to be "protected" from human beings. The functional view posits that human beings are a part of and are embedded in ecosystems, and the major role of conservation is to manage natural areas sustainably (also quoted in Baindur and Sarukkai 2010). This may clarify the purposes of conservation for both the categories, yet that does not take away from the fact that there is still a conceptual issue with conserving or managing "nature".

Both sustainability and preservation are fundamentally dependent on the concept of intrinsic value of nature. In a discussion about meanings of the word "intrinsic value", O'Neill (2003, p. 167) shows that the significance of some non-intrinsic properties is very important for the preservation movements. The rarity of a species or the biodiversity of any region cannot be invoked in these discourses, without using relational properties. Again, the human-ethical framework and the conceptual separation of the human from nature create the problem of establishing a connection, which could lead to an attribution of some extrinsic or intrinsic value for nature.

What normative concepts can be used to overcome the problem around nature and its value? One suggestion is that conservation can be a virtue practiced by human beings. Conservation in nature is a consequence of a more fundamental property of nature, namely various kinds of symmetries of nature (Baindur and Sarukkai 2010). The suggestion is that conservation has to arise from some other primitive property:

> What this suggests is that conservation in ecology need not be the primitive term but instead should arise as a consequence of some other property—either of nature or of humans. In a sense, such a view is already encoded in arguments that we should practice conservation because of our love of animals or of nature in general. What such an argument does is to make conservation a consequence of a prior proclivity—one that is guaranteed, because it is "natural". In what follows, we will argue that although it is important that conservation itself should not be a primitive term, such natural properties of love for nature are not necessarily the best ways to "derive" conservation (p. 17).

Thus from the above argument, a cultural legitimacy may be granted by making conservation a property of the human being. What then, could conservation be derived from that is a primitive property? In my earlier paper with Dr. Sarukkai, we argue that "conservation (either as a world view or as a practical habit) arises from a prior disposition to the idea of trusteeship" (Baindur and Sarukkai 2010, p. 18).

What is the implication of such an idea? One is that conservation itself is not necessarily a virtue, an "ought to be". On the other hand, it takes the form of a normative duty—an "ought-to-do". If we consider the conservation movement as an attempt to consolidate the idea that humans do have an obligation towards land and its occupants, the idea of trusteeship takes away the notion of ownership and replaces it with a sense of responsibility and obligation.

When we asked the research question, "whose lake is the Hebbal Lake?" we were first told that the lake was owned by the forest department. As migratory birds used the lake to rest, it was also a bird watcher's domain. When the lake was choked with weeds and silt in the mid-nineties, a local residents' welfare association (RWA) raised funds from the Norwegian government, desilted the lake, and created a small park on one shore. Further investigation revealed that in 2005, all lakes were meant to be controlled by the LDA. On the other hand, the horticulture department was in charge of fish culture and had contracted out the fishing rights to a fish stall owner. Further in time, the lake was leased out to a private hotel chain company to "develop" it. All these different stakeholders are related to the lake as controllers, owners, or as users. Every one of these controllers of the lake contested the right over the lake, ignoring the idea that they were entrusted with the lake, which was in some sense a public trust. Therefore, understanding conservation along the trajectory of trusteeship is related to conservation particularly within the Indian context can be beneficial:

> … trusteeship has an inbuilt relation to the idea of trust—this view makes it closer to a view of nature where our relationship with nature is based on 'trust' and not on 'confrontation' or on 'exploitation'. Trust is also to 'entrust' which is the spirit of trusteeship (p. 19).

The Indian word in Sanskrit for trusteeship is based on the word "to entrust"—*nyāsa*. The word has two sections: *ni,* which means "down", and *asa,* which means "to place", "to put", or "to throw". Trusteeship allows for an active engagement with nature and is a sustainable concept, as one does not conserve a trust for selfish ends. The advantage of such an ethical move is that the ethics remains social and is independent of problems of extension and concepts of pristine nature, etc.

Conservation thus gets defined in this context as the effort to preserve what is "entrusted" and the maintenance of it. In contrast to the idea of "stewardship", and also to the idea of a "stakeholder", a "trusteeship", land ethic would be the result of an ecological conscience, arising out of the responsibility to the land and other people as well.

10.6 Beings, Violence, and a Metaphysical Grounding for Ethics of Conservation

Popularly, Jaina and Sāṃkhya-Yoga schools of thought have an inclusive ethical credo of "do not cause pain" built into their philosophical theme of *ahiṃsā* that regard other sentient beings as capable of experience pain and suffering. *Ahiṃsā* in these two traditions is clearly articulated as the practice of restraint from violence. Though there is much literature on non-violence in Buddhist thought, it focuses on the virtue of compassion and will not be within the context of this section here.[10] According to a Jacobsen (1993), the principle of non-injury is based on a rejection of a natural tendency to live on other beings, in a world that is basically composed of beings dependent on feeding on one another. He points out "This discourse ascribes equal ultimate value to all living beings and embraces a transcendent beyond the realm of change" (p. 287).

[10]For more on Buddhist sources of environmentalism, see Kaza and Craft (2000).

Goodpaster (1978, p. v311) points out that it is important to distinguish between moral consideration and moral significance. The criteria for moral significance are comparative in nature: "... the comparative judgments of moral weight in cases of conflict". This means that if a particular resource is claimed by two sentient beings, one could distinguish whose claim is more significant. In case of the human and non-human, however, such a moral significance would depend on how much weightage each would have. However, the category of moral considerability would be an eligibility to be brought under a common ground (as we have seen in Chap. 3). However, the criteria for moral considerability (as distinguished from moral rights) could be the capacity to suffer or sentience. Such criteria would be useful in extending the moral umbrella over other non-human forms of life on earth. Within the framework of Western philosophy, sentience or life gets connected to moral considerability. And life here is sometimes articulated as a capacity to experience pain.

Moral considerability is also based on the idea of having interests. Debates among ethicists focus on what beings have an interest of their own, thereby allowing them to have moral considerability. Goodpaster (1978) writes:

> If one's conception of good is hedonistic in character, one's conception of a beneficiary quite naturally is restricted to beings that are capable of pleasure and pain. If pleasure or satisfaction is the only ultimate gift we have to give morally, then it is to be expected that only those equipped to receive such a gift will enter into our moral deliberation (p. 321).

In order to make sense of this concept of moral consideration, the idea of pleasure and pain has to be unpacked in Indian thought. Is it that only sentient beings that experience pain or pleasure are the recipient of our moral actions? The restriction of the experience of pain to life forms constrains the idea of an ethical relationship only to living beings. Yet again from a Western viewpoint, it becomes clear that ecological ethics is dependent on the recipient of the moral action. The ethical turn of the *mokṣa* explanation creates an argument analogous to the idea of "virtue".[11] The recipient of the effects of meritorious action is both the agent and the being towards which such action is directed. Here, I shall examine the idea of pain in Indian thought to clarify the issue a bit more.

What is the conceptualisation of pain and sentience in Indian thought? Most philosophies begin their explanation of their subject matter or aphorisms having stated that the explanations provide a means of putting an end to human misery and woes. The idea is not to seek pleasure but to avoid suffering. Matilal (2004), for instance, looks at the very important pain thesis or the doctrine of *duḥkha*. He seeks to analyse the statement: "The world is nothing but pain or suffering". Matilal (2004) argues against the idea that Indian philosophies are pessimistic. The pain thesis he proposes to interpret as a prescriptive statement.

> Suffering is a fact of life. From this empirical premise is developed, it seems, the thesis of universal '*duḥkha*'—where the term '*duḥkha*', I suggest, is no longer a descriptive term and

---

[11]One would say doing *dharma* is *good* for one's own state of spiritual evolution.

hence is hardly translatable as suffering. When '*duḥkha*' becomes an [Sic] universal predicate, its descriptive aspect I think becomes overshadowed by its evaluative function. '*Duḥkha*' in other words, is turned into, in addition, an evaluative expression (p. 17).

This statement forms a standard foundation for almost all of the ethical and moral injunctions that follow in traditional Indian thought. The first *sūtra*s of the *Sāṃkhya Kārika* state that the philosophical enquiry originates in the longing to put an end to the misery.[12] It is clear that ethical and moral practices are the steps to liberation, a way to escape from the suffering, or a way to the cessation of suffering.

10.6.1 Jaina Thought and Ahimsā

Jainism, in particular, places great emphasis on the practice of non-injury. Such a practice is based on the metaphysical understanding of a hylozoic universe filled with many diverse forms of life, and so any action is bound to cause harm to one living form or another. We have already seen in detail earlier the descriptions of the Jaina classification of life forms. We also have seen how the idea of *karma* and the need to attain *nirvāṇa* is the ideal goal of all life forms (see Chap. 5). To summarise our discussions from the earlier chapter, the Jaina world view of the universe as filled with myriad life forms adds to the danger of oneself committing violence that will endanger one's own spiritual progress: "It [*hiṃsā*] refers primarily to injuring oneself—to behaviour which inhibits the soul's ability to attain *mokṣa*" (Jaini 1998, p. 167). *Ahiṃsā,* in this school of thought, is not completely inspired by a "bio-centric" view of the world. The motivation for it is the doctrine of *karma,* the central obstacle to enlightenment. Violence as a form of passion attracts thicker karma that like an opaque material sticks to the soul binding it in the world of suffering.

Though the motivation for *ahiṃsā* may be salvation oriented, the practices of non-violence towards all forms of life are based on the fundamental premise that all life is sacred. The various practices of the Jaina community such as observing vows are all based on the ethical restraint of not causing harm to another living being. The lay people's vows, called *anuvrata,* are the practice of not harming all beings that have two senses or more. However, in the great vows (*mahāvrata*) followed by monks and nuns, the practice of restraint from violence is sterner. The monks make an effort to abstain from harming even the one-sense organisms. One could suggest that because of the foundational beliefs of the Jainas about *karma, ahiṃsā* is not strictly an ecological ethic and deny any possibility for this practice beyond its role in providing a means to salvation of its believers as *jīva*s. However, it can also be argued that belief in the connectedness of one's own well-being to the well-being of the environment within the Jaina principle of *ahiṃsā* allows it to be posited as a kind of anthropocentric environmental ethical principle. The only difference from the usual interpretations of anthropocentric ethics is that here well-being for humans

[12]Also for a related discussion, see Ghosh (1977, p. 1).

is the attainment of *nirvāṇa*. Wiley (2006) explains "Such concerns for that well-being of even the minute life forms, accompanied by voluntary restraints on the accumulation of processions and limiting the consumption of finite natural resources, accords well with a responsible environmental ethic" (pp. 46–47).

Another interesting idea in this perspective is the variability in the sensing of pain between one-sense beings and the many-sense beings. The one-sense beings have an indeterminate experience of pain; they are called *asaṃjñi*, without consciousness. On the other hand, the beings that have consciousness—*saṃjñi*—such as the two-sense beings, and others experience more pain. Such variability, Wiley (2006, p. 48) claims, can be used "... to create a guideline for priority of moral significance, with other beings ranked equally below them."

We find that when we treat *ahiṃsā* as a normative ethic, or as a virtue to be cultivated in human beings, there are many possibilities for an ecological ethics. With reference to the many practices of the Jaina community, one finds a reverence for life that results in a number of actions and a number of compassionate activities towards other beings that are guided by this overarching principle. Whether it is the establishment of animal hospitals and sanctuaries for injured animals or the fight for a ban on animal testing for drugs and cosmetics, the Jain community engages in real-world ecological practices related to animal rights.

The fundamental respect for the environment and nature springs from a self-awareness that any *vrata* holder must practice. Such an awareness that helps to keep them from deliberately injuring any living being translates into a thoughtful lifestyle that could be foundational in building other values of compassion or biocentric attitudes to nature. In the wider context of atrocities against nature that are thoughtlessly executed, remembering the Jain attitude to living beings and our relationship to them through the connectedness of the web of life may perhaps encourage us to abstain from needless violence.

10.7 *Ahiṃsā* in the Sāṃkhya-Yoga Philosophy

Sāṃkhya-Yoga philosophy stresses on *ahiṃsā* as among the most important principles to be followed by a moral person. The easiest understanding of the principle of *ahiṃsā* is that it is a practice of non-injury or in other words not harming any other being by body action, speech, or mind (*kāya, vāca, manasā*).

In Sāṃkhya philosophy, some insights about beings and pain create a new interpretation of sentience and pain. Here, the idea is derived—not from experience of pain by a being—by giving significance to concept of life-supporting function of nature. The concept of non-injury—*ahiṃsā*—one of the main tenets of the Sāṃkhya-Yoga ethics, is connected to the idea of not causing pain to living beings but different from the Jaina interpretation.

Ahiṃsā is of course popularly understood as linked to the idea of experiencing pain which in turn is dependent on a certain view of sentience. While the ordinary view of "sentience" is given by the very important property of a being—having "life"—there is a unique interpretation of this concept that we find in the Sāṃkhya philosophy. The commentator Aniruddha, cited by Larson, explains the problems of the types of beings and claims they cannot be restricted to the human condition alone: "Trees, bushes and plants and like-wise are abodes of experience for an experience". He also explains that meritorious behaviour however is not possible for them like people due to bodily conditions (as elaborated in the scriptures).

Sāṃkhya admits 14 types of sentient life in the phenomenal world called *bhautikasarga.* Aniruddha classifies these bodies into three types: "(a) bodies that act (*karmadeha*), (b) bodies that experience (*bhogadeha*), and (c) bodies that both act and experience (*ubhayadeha*)".[13] Besides the plants, animals, and human beings, there are eight celestial realms of beings that are specially referred to as *māhātmya śariras* (great-bodied beings). While the other life forms (humans, plants, and animals) are embodied according to the merits and demerits following their karma, these celestials have bodies created by will. Due to the nature of their cosmic duties (*dharma*), their transmigrations occur among stars, planets, and other cosmic elements.[14] This description makes almost the whole of the phenomenal world sentient in some intrinsic way.

The argument in the Sāṃkhya scheme of things is that any action of any being is inherently violent in some way or the other. Called *upahanana,* any action in *prakṛti* yields pain (*duḥkha*) because in every action, there is always some form of violence. Due to the admixture of *guṇa*s (particularly *rajas*), a "pure act" is never possible. Therefore, every act has to be compensated by other meritorious actions or restraint. Even violent acts enjoined by *dharma* have to be compensated. Vaṃśīdhara Miśra (quoted in Larson and Bhattacharya 1987, pp. 433–434) clarifies this by of explaining the seemingly contradictory Vedic injunctions to sacrifice animals on one hand and to hurt no living being on the other: "Violence of all kinds invariably gives rise to the ill results to the person concerned. That is why the victorious *Kśatriya* heroes of the *Mahābhārata* had to perform expiatory rites (*prayascitta*)". Perhaps this presence of violence in almost all acts is why *Yoga* advocates minimum interference in *prakṛti* and withdrawal as necessary for liberation (p. 297). Jacobsen (1993, p. 287) considers that the dependency of organisms on each other is a part of the *hiṃsā,* and therefore, the ethics of non-injury places the emphasis on withdrawal from this cycle of violence. He points out that the purpose of *ahiṃsā* as oriented towards "salvation of souls", makes it is quite difficult to adopt such a principle of total withdrawal into an ecological ethics directly. However, if we reinterpret the notion of *ahiṃsā* pragmatically, independent of a binary (as it was not intended for the practice of common people in its extreme form), it is possible to invoke the idea of "protecting beings that support other beings".

[13]"Aniruddha: on the problems of the beings," summarised by Gerald James Larson (1987, p. 368).

[14]I am grateful to the conversations I had about this with Siddhartha Arya, a traditional Saṁkhya student. Also see a summary of the beings descriptions in Larson and Bhattacharya (1987, pp. 59–60).

Ahiṃsā in its complete form is almost impossible to cultivate while leading an ordinary non-yogic life. However, the aim of all human beings would be to reduce the *hiṃsā* (injury). A gradation of *hiṃsā* (violence) is possible in Indian thought based on a unique insight into the mutual relationship of the beings to each other. Sāṃkhya thought states that the world consists of beings that are in a relationship of mutual dependency and support. The "interdependency" (*parasparārthatva*) of the divine beings, the human beings, and the animal and the plant beings is also elaborated in the *Yuktidīpikā* and the *Jayamangala* [traditional commentaries] on Sk 15 [*Sāṃkhya Karika,15*] (from Jacobsen 1999, p. 328). In Sanskrit, this relationship is called *āśritabhāva* when the being is dependent or being sheltered and *aśrayabhāva* when the being supports other or gives shelter. There is a very important conceptualisation that could be understood from these two ways of looking at creation. Every created being that is a part of the phenomenal world exists in a way that it supports the existence of other beings. From the typical notion of an embodied being occupying a world that is a mere empty field of experience or life for it, this notion of dependency and support posits the idea of a series of beings that are embodied in *prakṛti* and themselves as *prakṛti* en-world other beings. The similarity with an ecological perspective cannot be missed.[15] The worlds and beings do not continuously get larger and larger, but end with those final beings that provide support as cosmic beings that are called *prakṛtilīna* (embodied as *prakṛti*). The body of such beings is one with the entire universe or the natural world. Just as beings have parts of the body, these nature-embodied beings have cosmic parts such as the sun and the planets. *Sāṃkhyakārikā* gives examples of the cosmic beings such as *Vaiśvānara* or *Hiraṇyagarba* mentioned in the Vedas. To understand creation as a being is to attribute a certain kind of relationship between all beings. This beingness does not presuppose a naïve understanding of sentience. Rather, it seems to refer to the universe as a set of internally relational beings at different levels of existence, each embodied and at the same time en-worlding other beings, all bound up in the eternal process of transmigration. The gradation of violence is based on this, both in the human and other realms. To cut a tree that is an *aśraya* (shelter) for a number of beings is worse than say cutting a branch and injuring a tree. Socially killing a person with dependents (who have no other shelter) is treated as a more immoral act than killing a person with no dependents.[16]

Within this important concept is hidden the idea of not only an ethics that takes into account non-human nature, in the form of animals and sentient beings, but also the land, water, air, and all the forms of habitats and ecosystems. As we have seen in Chap. 4, *pañcabhūta*s that are often referred to as the five elements are very important in the Indian traditions of thought. These five elements are currently

[15]Attribution of sentience to the environment may be problematic, if sentience is interpreted according to the idea of experiencing pain and pleasure; from the *Sāṃkhya* viewpoint, however, the experience is separated from the awareness of it. All beings may experience pain, but based on their category of life forms, they may not be aware of their experience. This is somewhat similar to the Jaina view we have seen earlier.

[16]Killing a king therefore is considered an intense act of *hiṃsa* as he is the shelter for many people.

being viewed as commodities, rather than as parts of a larger cosmic being that supports our existence. Just as we do not view the parts of our home as economic resource to be sold, similarly we do not treat the cosmic nature—*prakṛti*—as a commodity (Rao 2000, p. 37).

It is important to remember that the *Sāṃkhyakārika* is not talking about a mere food chain type of welfare and dependency when it refers to *aśrayabhāva*. Conservation in this framework does not require us to reinterpret our relationship with the non-human world in radical ways. We could just use this concept to understand that we as the collective of human beings in today's world have to function as the being that has the *aśrayabhāva*. There might have been a time when human beings were supported by various other beings. The relationship of human beings to other beings has slowly changed from being dependent (*āśrita*) on nature to that of being a shelter (*aśraya*). As a collective being, the human population provides sustenance to many beings in nature. The *dharma* or duty of a being that gives shelter is contrary to the act of destroying the sheltered. This act of giving shelter is a powerful moral ethic in Indian thought. The protector, who promises shelter to a dependent, is often willing to go to great extents to fulfil the act of protection, even at great personal sacrifice.[17]

From this perspective, in today's world, due to the dominant place occupied by human beings, they are no longer allowing other beings "to be". This, according to Sāṃkhya-Yoga philosophy, would be *hiṃsā*. Destruction of "shelters" in the form of habitats is a worse form of violence that involves killing individual animals. To destroy that which supports of existence of other beings, through the actions of one's body, mind, and speech, is *hiṃsā*. Even to plan, speak of, and create policy that encourages such destruction would be morally wrong.

From the insights above, two possible moral action norms can be derived. One is that since humans are the shelter for other beings, it becomes the duty of humans to work to protect their lives by whatever means, even at the cost of some human welfare or profit. The element of sacrifice is important here, especially when it involves sacrifice of economic benefits. This is a proactive stand to take on protecting other beings and habitats collectively as a human race. Secondly, human destruction of nature is morally reprehensible. This is a "restraint" kind of ethics wherein humans intentionally should not cause harm to other shelterers. The destruction of the "beings" such as those that provide shelter to many more beings (ecologically one could call them habitat-beings) is a grade worse than say killing an animal for food. In the definition of "greater good", Sāṃkhya provides the necessary value for judgment. For example, destroying trees on which many beings live, clearing forests, or draining water bodies would be morally significant than say eating meat. Expressed in modern terms, the ecological impact of violence would have to be taken into account in every human endeavour to modify nature and

---

[17]Stories of Śibi, the great king who gave his flesh for the food of a hawk to replace the meat of a dove that seeks his shelter, is recounted in the *Mahābhārata* and popular legends.

natural habitats. Choosing restraint and welfare of other beings (especially those that come under the *aśraya* category) over-violence and self-profit, would be a natural framework for an ecological ethic.

10.7.1 Land Ethic in Practice: The Bisnois

The earlier discussions of landscape are focused on secular traditions and landscape as nature. The difference between these traditions of thought and the imagination of scared nature is that here land does not have to be transcendent or divinely originated or blessed in order to be respected, protected, or understood. Many indigenous communities also envision nature as sacred. Besides these, there are a few narratives of nature that are ethically related through prescriptive principles. These may not be included under the narrower definitions of religious beliefs but fall under somewhat broader and ambiguous category of lifestyle practices or traditions. Any discussion of relationship between human beings and conservation in India would be incomplete without recounting the practices of a sect called the Bishnois.[18] It is said that in 1730 AD, 363 Bishnois lead by a woman Amrita Devi Bishnoi sacrificed their lives to oppose the cutting of *khejari* trees by palace workers in Khejarli village of Jodhpur. She hugged the tree and said, "*sir santhe runkh rahe to bhi sasto jan*", meaning that a chopped head is cheaper than a tree (Brockman and Pichler 2004, pp. 17–18). By the time the king rushed into stop the killing, his soldiers had killed all the people who had stepped up to embrace their trees. The repentant king then passed a law that forbade the cutting of trees and hunting in that region that is upheld by this community even today. The shrine I visited was the place where the final ceremonies were ritually conducted for the people who died along with the trees. The shrine, like the unassuming people who pray, is small with an idol of their teacher. The Bishnois do not believe in idols, and the statue is a mere marker of respect. There are a few paintings depicting Amrita Devi's sacrifice and a large platform for gatherings. Many such "*saka*s" or places of sacrifice are marked with names of those who gave their life for trees. There are a few narratives and books available in their local language about these movements of extreme sacrifice where human beings gave up their life to protect trees or animals. Brockman and Pichler (2004, pp. 20–21) recount the earliest instance of self-sacrifice of how in 1604 AD, and two Bishnoi women from Ramsari village, Karma and Gora, sacrificed their lives in an effort to prevent the felling of *khejri* (*Prosopis cineraria*) trees.

Though, a separate sect and religious community who worship their teacher Jambheṣhwar (considered an avatar of Viṣnu), they are not followers of Vaiṣṇavism,

[18]There is very little academic literature on the Bishnois whose narratives find more presence in folklore and media reportage. Most of my work references a book by Brockmann and Pichler (2004) and also personal narratives gathered during a field visit to Jodhpur district of Rajasthan where the great sacrifice of Amrita Devi is said to have occurred.

often the community are wrongly called "Viashnavs" or "Bishnavs". They are actually called Bishnois because of the 29 precepts they follow—*bish* (twenty) and *noi* (nine). Central to the sectarian practice of Bishnois is a relationship with nature that is based on 29 principles taught by their sect's teacher. In practice, since the early fifteenth century, these prescriptive principles include injunctions against the cutting of a live tree or harming any living beings or using indigo-dyed clothes (the dye is made from crushing the living plant). All of these injunctions are based on fundamental respect for the delicate balance between the land and its inhabitants. The Bishnois believe that trees and other sentient beings are members of their own kin, and any harm caused to them is similar to the harm caused to a member of their own family. This has resulted in the areas occupied by Bishnois being rich in flora and fauna. This resource-rich area attracts exploiters of nature, and the Bishnois often stand as custodians or guardians of their environment. Protecting trees and wildlife in accordance with the do-no-harm credo, the community engages with the system of passive resistance, a precursor to the modern tree-hugging movement. There are many local stories of Bishnois taking on powerful poachers or politicians and saving sacred life by sheer sacrifice.

During my visit to the shrine erected at the final cremation site of these martyrs in December 2012, a local Bishnoi recounted for me the importance of the *kejari* tree in their life and in the desert. He pointed out that the sturdy tree was important to keep out the desert and to provide them with fodder, and it was also a delicious source of vegetable for a much-liked dish.

Unlike other narratives of nature and philosophical traditions, the practices of the Bishnois are still in line with their principles. As Brockman and Pichler (2004, p. 48) point out, though the ethical values of these people are derived from religious or cultural sources, the by-product of these practices is the richness of flora and fauna. As I travelled through their land, I saw that the blackbuck and *nilgai* were allowed to eat from Bishnoi fields and the people did not regard them as pests or intruders. The environmental success of their lifestyle is evident. *Kejari* trees and greenery line their roads and fields. Interspersed near their shrines and holy gathering places are small ponds and lakes managed locally for wildlife to access water. A wildlife sanctuary has been established bordering their lands. They are fiercely protective of their wildlife and trees and still continue to oppose tree cutting and poaching. A recent transgression of sorts occurred in 1998 when a popular cine star was involved in hunting down a blackbuck in their area. The Bishnois who complained stood firm as witnesses against him and are fighting a long and hard battle even now to have him punished legally for hunting wildlife.[19] One would without hesitation call the Bishnois as the earliest environmentalists or eco-warriors of our times.

The Bishnoi practices are not mere lip service or pretensions of a theoretical claim to philosophical concepts of environmental protection. My visit to this land of Bishnois showed me how environmental behaviour can become embedded in the everyday lives of people, withstanding even the modern world of machines and gadgets.

[19]For more on this case, see Brockman and Pichler (2004, pp. 8–16).

References

Baindur, M., & Sarukkai, S. (2010). The possibility of conservation: Some philosophical questions. *Indian Journal of Ecocriticism, 3*, 9–21.

Baindur, M. (2014). Bangalore lake story: Reflections on the spirit of a place. *Journal of Cultural Geography, 31*(1), 32–56. doi:10.1080/08873631.2013.873296.

Balasubramanium, R. (2000). Ecophilosophy in the Indian perspective. In L. Surendra, K. Schindler, & P. Ramaswamy (Eds.), *Stories they tell*. Madras: Earthworm Books.

Bilimoria, P. (2007). Introduction to part A: Early Indian Ethics—Vedas to the Gītā; dharma, rites to 'Right'. In P. Bilimoria, J. Prabhu & R. Sharma (Eds.), *Indian ethics: Classical traditions and contemporary challenges* (Vol. 1, pp. 33–56). New Delhi: Oxford University Press.

Brockmann, H., & Pichler, R. (2004). *Paving the way for peace: Living philosophies of Bishnois and Jains*. New Delhi: D.K. Publishers and Distributers Pvt. Ltd.

Callicott, J. B., Crowder, L. B., & Mumford, K. (1999). Current normative concepts in conservation. *Conservation Biology, 13*(1), 22–35.

Chapple, C. K. (2007). Action oriented morality in Hinduism. In P. Bilimoria, J. Prabhu, & R. Sharma (Eds.), *Indian ethics: Classical traditions and contemporary challenges* (Vol. 1, pp. 351–363). New Delhi: Oxford University Press.

Dasgupta, S. (1922). *A history of Indian philosophy* (Vol. 1). New Delhi: Motilal Banarsidass Publishers Private Limited.

Deutsch, E. (1989). A metaphysical grounding for natural reverence: East-West. In J. B. Callicott and R. T. Ames (Eds.), *Nature in Asian Traditions of Thought*. Albany: State University of New York.

Goodpaster, K. E. (1978). On being morally considerable. *The Journal of Philosophy, 75*(6), 308–325.

Gambirananda, Swami. (Trans.). (1957). Kaṭha Upaniṣad. In *Eight Upanishads: With the commentary of Śaṅkarācārya* (Vol. I). Calcutta: Advaita Press.

Ghosh, J. (Ed.). (1977). Introduction. In A. Swami Hariharananda (commentary and translation) *'Sāṁkhya—sutras' of Panchashika and Sāṁkhyakhyatattvaloka*. New Delhi: Motilal Banarsidass.

Houtan, K. S. V. (2005). Conservation as virtue: A scientific and social process for conservation ethics. *Conservation Biology, 20*(5), 1367–1372.

Jacobsen, K. A. (1993). Restrained persons, places transformed: Relation to place in the Saṃkhya-Yoga traditions. In P. B. Singh Rana (Ed.), *Environmental ethics: Discourses and cultural traditions, a festschrift to Arne Naess* (Vol. 39, Issue No. (1–2), pp. 141–49). Also published in *The National Geographical Journal of India* .

Jacobsen, K. A. (1999). *Prakṛti in Sāṁkhya–Yoga: Material Principle, Religious Experience, Ethical Implications*. New Delhi: Motilal Banarsidass Publishers Pvt. Limited.

Jaini, P. S. (1998). *The Jaina path of purification*. New Delhi: Motilal Banarsidass.

Kaza, S., & Craft, K. (Eds.). (2000). *Dharma rain: Sources of buddhist environmentalism*. Boston: Shambhala Publications.

Kirkpatrick, A. (2001). Exploring differing approaches to environmental responsibility. In *Environs: The Student Journal of Environmental Studies* (Vol. 1(1)). Sackville, New Brunswick: Mount Allison University Canada.

Larson, G. J. (1987). 'Conceptual resources' in South Asia for 'environmental ethics' or the fly is still alive and well in the bottle. *Philosophy East and West, 37*(2), 150–159.

Larson, G. J., & Bhattacharya, R. S. (Eds.), (1987). *Sāṁkhya: A dualist tradition in Indian philosophy*, Encyclopedia of Indian Philosophies (Vol. 4). Delhi: Motilal Banarsidass Publishers.

Leopold, A. (2006). The Land Ethic. In A. Light & R. Holmes III (Eds.), *Environmental ethics: An anthology*, (pp.38–46). Malden: Blackwell Publishing. Earlier published in Leopold, A. (1949). *A Sand County almanac and sketches here and there*. New York: Oxford University Press.

Light, A. & Holmes, R., III. (2006). Introduction: Ethics and environmental ethics. In A. Light & R. Holmes III (Eds.), *Environmental ethics: An anthology* (pp. 1–13). Malden: Blackwell Publishing.

Marshall, P. (1992). *Nature's web: Rethinking our place on earth*. London: Simon and Schuster Ltd.

Matilal, B. K. (2004). Duḥkha. In *Logical and ethical issues: An essay on Indian philosophy of religion*. Reprint edition. (1982), New Delhi: Chronicle Books.

Mohanty, J. N. (2000). *Classical Indian Philosophy*. Maryland: Rowman and Littlefield Publishers Inc.

Mohanty, J. N. (2007). Dharma, imperatives and tradition: Toward an Indian theory of moral action. In P. Bilimoria, J. Prabhu, & R. Sharma (Eds.), *Indian ethics: Classical traditions and contemporary challenges* (Vol. 1, pp. 57–78). New Delhi: Oxford University Press.

Nelson, L. E. (2000).The dualism of Nondualism: Advaita Vedanta and the irrelevance of nature. In L. E. Nelson, (Ed.), *Purifying the earthly body of god: Religion and ecology in Hindu India* (pp. 61–88). New Delhi: D K Printworld (P) Ltd.

Seshagiri Rao, K. L. (2000). The five great elements. In Chapple, C. K., & Tucker, M. E. (Eds.), *Hinduism and ecology: The intersection of earth, sky, and water* (pp. 23–37). New Delhi: Oxford University Press.

Smith, R. L. (1976). Ecological genesis of endangered species: The philosophy of preservation. *Annual Review of Ecological Systems*, *7*, 33–55.

Soulé, M. E. (1985). What is conservation biology? *BioScience, The Biological Diversity Crisis*, *35*(11), 727–734.

Wiley, K. L. (2006). Nature of nature: Jain perspectives on the natural world. In *Jainism and ecology: Nonviolence in the web of life*. New Delhi: Motilal Banarasidass Publishers.

Chapter 11
Creating New Paradigms of Understanding: Action and Ecology

Abstract As the final concluding chapter of my book, I have focused on certain significant implications of the conceptualisations of nature in Indian thought on moral action and conservation. The applicability of these interpretations for creating a framework of ecological ethics is analysed. The conclusion of an "eco-moral action" based framework for ecological ethics then places the themes discussed earlier into a modern context for conservation and other ecologically relevant themes.

Keywords Ecological ethics · Conservation · Human–nature relationships · Implications of nature as a concept

11.1 Conservation

An issue around conservation in practice is the gap between people's understanding of nature and the scientific expert understanding of the world. As a part of the lay–expert divide, general people are often seen as ignorant of the "science" that informs "experts" who determine the policy and management of conservation. The gap becomes critical when the cultural values accorded to nature by the people differ from the values accorded by conservation biology or the discourse of "conservation" in science.

According to O'Neill (2003), such environmental evaluations or parameters cannot be included in any environmental ethics framework that includes non-instrumental or intrinsic value. In other words, if a value is relational and evaluated in comparison with other objects, it cannot be of intrinsic value. Can an object be evaluated in such a way that it depends only on its intrinsic value? O'Neill (2003) concludes this discussion by stating that meta-ethical questions may not be required by an environmental ethic as much as normative and applied concepts.

Conservation ethics bases itself on the idea that "nature" should be conserved, and at the heart of this is the idea that it should be untouched by humans. There is

M. Baindur, *Nature in Indian Philosophy and Cultural Traditions*,
Sophia Studies in Cross-cultural Philosophy of Traditions and Cultures 12,
DOI 10.1007/978-81-322-2358-0_11

already a contradiction of sorts. On one hand, nature has to be left to its own devices, without human intervention, but yet it has to be "managed" by conservationists as much as possible. In practice, conservation is more about spreading awareness among people who live close to these demarcated areas. In an attempt to conflate science and values, environmental education programmes attempt to teach facts about conservation to local people. However, facts alone do not help to motivate conservation, claims Trudgill (2001, p. 680).

Conservation therefore includes many human activities, such as managing species population by protection, creating exclusive preserves, restoring habitat and degraded ecosystems, promoting sustainable uses of nature, and measuring or evaluating various parameters that indicate the "natural well-being." In this activity of conservation, ecologists, and conservation biologists increasingly see ethics as empirical ethics.

We have seen earlier how the idea of conservation especially as related to nature itself is problematic. Is conservation a property? Is it an action, a process? Is it an ethical stance towards nature? Is it management strategy to keep nature untouched like a historical artefact? The following section of the chapter discusses the idea of conservation and Indian theories of nature to uncover an approach to conservation as a moral duty of human beings who co-constitute nature. One alternative is to understand conservation through the concept of trusteeship as advocated by Gandhi. I reinterpret this idea to see how, from a "conservation of natural resources" perspective, this can form a foundation for a moral eco-ethic. Another alternative is to use the already popular idea of *ahiṃsā*, or non-injury, from an ecological context. To create a context of non-injury, that works beyond the idea of a one-to-one personal violence.

11.2 On the Concept of Nature

Ecological ethics in Indian thought is based on the premise that human beings are intrinsically related to all the other created existents in the universe. At the outset, this seems a rather sweeping view of oneness, but as we have seen in earlier chapters, this relationship is both moral and metaphysical at the same time. As mentioned earlier, a large number of thinkers in environmental philosophy look for a common ground or a framework, within which non-human components of the universe could be morally significant or be morally "considerable". The common ground would place human beings on an equal footing with the rest of nature, establishing interconnectedness or ways of relating to nature that would lead to an ethical recognition of the need for conservation.

The environmental crisis itself has been articulated as a problem of "nature", rather than that of the human being. There have been attempts at solving the "problem" from two angles conceptually. The issues and concepts that are discussed in this chapter and the next are based on attempts to create a framework for of eco-ethical action; firstly, to see how one can highlight phenomena of relationships between human beings and the rest of the environment as envisioned in

Indian thought, and secondly, it is an attempt to re-describe conservation itself from Indian perspectives.

During my lake study (see Chap. 10 for discussions), I found that the lake was subject to natural variations of water content in the past. While the water would dry up in the summer months, during the monsoons, it would overflow into adjoining fields and wetlands. People saw the "*keré*" (lake) as something that varied with the seasons. However, for the urban developer, the conceptualisation of the lake is unvarying over all seasons. The Western concept of nature as non-human fails to address the question about newer "natural" objects created in the human world. For instance, from my field study of the urban lake questions such as "Is the lake natural? Or is it an artefact?" become points of ambiguity.

Extending this argument to the concept of nature itself, we can say that while nature is unpredictable, unstable, and constantly changing, a certain conceptualisation of nature is a constant. Parks and gardens package this as "wilderness" and "nature" for the urban dweller by the process of landscaping.

As earlier discussed earlier in Chap. 3, the natural and the artificial are problematic categories. If seen as *prakṛti*, the lake is still nature, and the rapid rate of conversion of a wetland into dry areas for buildings and recreational complexes is caused by human beings. By creating barriers to the natural inflow and outflow areas of a "*keré*", we can say that we are changing the *dharma* of the water body, thereby its function too. And it is important to remember the causal arguments of the Sāṃkhya philosophy. In this case, the *nimitta kāraṇa* are the human beings (such as the people in authority, policy makers, and the private company), while concomitant conditions (*sahakāri śakti*) such as urbanisation, pollution, and development of real estate are also present. I am not claiming here that the Sāṃkhya viewpoint offers a solution, but I suggest that it helps us to include many more factors into the problem, giving us a richer detail than say a "cost/benefit ecological economics" analysis or a "preserved/degraded" ecological analysis. The other question that can be asked from the Sāṃkhya viewpoint is whether all the permanent buildings set up on the lake shore under the lake development project are a "milk to curds" type of change or a "mud to unbaked pot" kind of change.

The policy to restore the lakes to an urban island of "greenery" seems to ignore the everyday realities of the daily interactions of the various people who are connected to the lake. Instead, there is the dominant influence of a large-scale conceptual model of conservation, based on the idea of "clean and green" that seems to be in direct conflict with the idea of public or functional use. There exists a lack of clarity in such a framework for a philosophy of conservation, where eco-ethical actions are different from our everyday interaction with nature and the world around us. In the action theory of Indian philosophy (which is well articulated in Advaita school of thought), I believe there is a conceptual understanding of moral ethics, that is beyond the understanding of right and wrong in ways we understand them today. Those actions that benefit the environment in the long-term could be considered as moral actions, and those that provide us with short-term economic benefits, pleasures, and luxuries could be considered immoral action. This new paradigm given by this theory could have us understand action as eco-moral action or as eco-immoral action.

The conceptualisation of *prakṛti* as nature, as I have suggested earlier, creates a world view where the rate of modification of nature into refined objects becomes central to the understanding of conservation. There I posit that conservation in this perspective is a slowing down of the change, not necessarily eliminating it. So alternatively, conservation–action can also be interpreted as *dharma* or a property of the human beings towards other beings in the context of caring, through an attitude of trusteeship or the act of giving shelter to other beings.

We have seen earlier theories of moral actions deeply linked to the idea of liberation or sometimes the attainment of good *karma* oriented towards some sort of soteriological goal. Critiques of the moral action theory in India have insisted on an ethical motive that is not so directly linked to the transcendental goals of a human being. However, within the theory of action in the *Bhagavadgītā*, there may be some useful concepts that relate to an ecological stance.

11.3 Disinterested Action: Non-consumption as Ethical Action

Following Larson (1987)'s suggestion that any work on concepts for environmental philosophy has to overcome these earlier mentioned fallacies, I propose to invoke Deutsch (1989)'s idea of disinterested action as an insight for a new understanding of the problems of ecological crisis. The foremost problem in this crisis seems to be the problem of conserving nature, both in the form of natural resources that we are dependent on, and the non-human parts of a natural world. I think that action that is *karma-yoga* or "disinterested action" can be derived from Advaita philosophy using a very different perspective, taking into account the interpretation of "action" as embodied action in the world for ecological ethics. Rather than the interpretation as an abstract principle that is about being 'unattached to the fruit of action', *niśkāma* can be interpreted as 'restraint' in this view. Overall, the message of reducing "consumption" is clear in the various Indian philosophies such as Advaita, Sāṁkhya-Yoga and Jainism.

Focusing on desire, rather than liberation, overcomes at least one of the fallacies that Larson (1987) raises—the fallacy of symmetry. The human capacity to desire has not changed over time, though what we want and how much of it we want have changed in the modern age. Human desire that is a core meta-problem of sustainability, conservation, and the ecological crisis of environmental destruction is not only about the individual, but also includes collective desires. As social congregations of people, we embody collective goals, ambitions, and desires. Measuring human desire against human need will actually give us a very good idea of our own conception of ourselves as morally responsible for actions on the earth. Perhaps, in understanding what we want and what we need, the fallacy of the sovereignty of the subject can be overcome. This is not a new concept. Many activist organisations have recognised that blatant consumerism has led to exhaustion of resources and created stress on the requirements natural world is left around us. Increasingly, energy-dependent lifestyles are set to leave a large "ecological footprint." The effect of environmental destruction on a community and its surrounding environment in Plachimada illustrates this point.[1] Most people are quick to accuse the multinational soft drink company for the environmental crime of water contamination and overusage of resources. But, at an ethical level, every consumer who takes pleasure in that brand of soft drink has played out his or her desire of the senses. If nobody desires the soft drink and drinks just water, is it not likely that the environmental destruction would not have taken place?

Indiscriminate desire is ethically wrong in the environmental context. In my own understanding, it is eco-ethically immoral. Actions performed in order to satisfy desire are "not right". They are not wrong in the sense of "evil", but wrong in the sense of leading away from both proximate and ultimate purposes of human life.

Going back to the illustration of the lake development project again, one of the "development" activities was the planning of a food court on the lake shore. People would also take a boat out to a "floating restaurant" and have food, entertainment programmes, and parties. Among the issues raised by the environmental activists was that there was no need to set up such an elaborate "fun area" at tremendous environmental and social cost on water or near a lake. They claimed that since a city such as Bangalore already had many places for people to enjoy food and also many entertainment halls, there was no reason why the lake should be considered for this type of a "development". Activists perceived this as exploitation of the lake area by the private company, as it was a well-known hotel chain group. They claimed that the company was using a "public space" as an excuse to set up its hotel activity, without paying for real estate costs.

It is true that soft terms such as "aspiration" or such as "development" are very much in the discourse of this "disease of desire" that currently seems to have acquired an ethically permissible existence. While environmental thinkers focus on reverence to nature, relationship to nature, metaphysical oneness of humans with nature, not many focus on the moral and ethical foundation of many Eastern philosophies—the reduction of want and the restraint of sense-pleasure and emphasis on a life that moves towards simplicity and unattached action.

In the case of the lake, the discourse of "developing the lake as a green and natural area, for people to enjoy", was prominent. The "developed" lake thus begins to embody values that are given by different discourses of aesthetic values and the multitude of parameters that represent these values. The physical transformation of the lake, from a wetland marsh into a drinking water reservoir and on its way to being a component of a public park, follows this conceptualisation of what the lake is, and how it should be managed. Though ultimately, the planners tended to look at trade-offs between these various values, it became clear that the functional values of lakes, which involve direct usage of resource such as water conservation or fishing, are of lesser importance than the lake forming a visual and aesthetic backdrop for

[1]For details of this environmental incident in the state of Kerala, see Bijoy (2006)

recreation, in the form of a fun world with eateries, stalls, and shopping centres. As one of the participants in the research project said, "It is about how to consume the lake by paying money."

Though very complex and subtle, it is also very easy to see that human action towards nature is linked to the everyday human action, caused by the desire to consume. If collective human action is guided by a satisfaction of pleasure—an indulgence of senses—then such a desire-based action is to be avoided. Again, it is easy to argue that the line between good and pleasure is very difficult to recognise, as is the line between need and greed, or between necessity and luxury. However, in case of the ecological crisis, it is clear that need and greed are socially, geopolitically, and culturally dictated. Despite this, one can insist that across cultures, it is possible to recognise the profanity of excess, and the overexploitation of nature is not so hidden from common-sense morality.

In today's world, there is so much emphasis on nature as pristine that most environmental philosophers forget that everything that we have around us, so-called artificial, or all manmade objects draw their primary resource from nature. So the modern person tends to see no "nature" in a laptop for instance. It comes to be of no surprise than that we continue to lead lifestyles that exhaust our resources and still wonder why there has to be a crisis. The connection between a waterhole drying up in a jungle and the use of a car everyday seems to be missing in the mind of the common people. The connection between my actions everyday where I consume resources is not considered from an eco-ethical viewpoint as much as my direct involvement in some action labelled as "eco-friendly", such as signing a pledge to "Save the Tiger".

When we interviewed two software engineers walking in the park on the shores of the lake, they were very happy with the idea of taking a boat down to a restaurant and having a cup of coffee. They conceptualised the floating restaurant as a calm retreat with a pleasure-giving coffee break. However, they were unable to gauge the complexities of the resources that were required for such a project or the effect on the water or the wild birds around them. On the other hand, the view of an office bearer of the eco-wing of a local resident welfare association (incidentally named Thoreau foundation) that was involved with the upkeep of the lake was that the silt islands in the lake should be populated with deer, to make the lake "more natural". From a naïve viewpoint, the second suggestion seems perhaps better suited to a conservation effort. But on being unpacked, both are desires of the human to enjoy something, excesses that we may well do without on a lake shore.

Going by the philosophy of Advaita and the theory of moral action, every action performed is a moral action directed towards restraint or directed towards sense pleasure. It is this concept that underlies the slogan of conservation "to reduce, recycle, reuse." To reduce consumption of resources, we need to reduce desire that is the root cause of consumption. The focus of a philosophy of conservation is human nature and not nature itself. Gandhi emphasises on this very idea of restraint of desire when he suggests that the "Earth provides enough to satisfy every man's need, but not every man's greed."

Guha (1989) explains that the roots of global ecological problems lie in the disparate sharing of resources. The industrialised countries and the elite of the third world consume more resources than they need. He suggests that the solution for the ecological crisis lies in the West adopting alternate political and economic structures and also changing some of their cultural values. Guha (1989) also argues that the attempt of deep ecologists to insist that intervention in nature should be guided by principles of "conserving biotic integrity" would have harmful effects in third world countries. He suggests this is because the dichotomy of a bio-centric versus an anthropocentric viewpoint is of little use in the third world. Particularly contrasting India with countries such as the United States, he insists that in developing countries such as India, creating pockets of wilderness would actually displace the agrarian communities who have lived in interaction with nature. Rather, his suggestion is that the ethics of restraint be adopted by the West: "The expansionist character of modern Western man will have to give way to an ethic of renunciation and self-limitation, in which spiritual and communal values play an increasing role in sustaining social life" (p. 249).

11.4 The Relational View of Ecological Ethics

There is a need to understand the category of an ecological ethics that is different from the domains of environmental ethics. While environmental ethics deals with appropriate management of the environment as perceived as natural resources and as sustainable for human use, ecological ethics is about the moral relationships between human beings and nature. This view of an ethical response to the ecological problem is broader and does not reduce our experience of nature to one particular view of nature. The challenge therefore is to bridge traditional accounts of nature with current prevalent concepts of nature, and this can be achieved through synthesis rather than through positivist shifts. The richness encountered by human beings as nature is diversely captured in many cultures of thinking and speaking about nature. To engage with these streams of thought would certainly yield rich dividends for the ecological cause.

For instance, Berkes et al. (1998) suggest that "ecosystem"-like notions are found in traditional cultures including concepts of bio-regionalism and "sense of place". They suggest that many indigenous peoples have words in local languages that get translated as "land", which often refers to a broader and richer category that is inclusive of the human. Such traditional understandings of the "ecosystem" move away from a positivist mechanical perception of nature towards a more organic interpretation of biological networks, inclusive of human beings, and their experiences. Every human being experiences nature as a place. The phenomenon of place is the experience of a space that has somewhat absorbed into it narratives and meaning that people ascribe to it. These narratives do not exist in the mere imagination of the people, but we find them represented in tangible elements and real fragments of the physical and material. Casey (2001) claims that place not only provides a location of a "where" things happen, but also provides the "how" and the "when" of one thing relating to another. These meanings are vested in physical matter. For example, in the case of Hebbal Lake that we looked at in the previous chapter, these meanings are vested in the discarded torn nets of the fisher folk or in the flat stones used by the dhobi or the favourite footpath with worn-out grass used by the bird watchers, or perhaps represented more powerfully in elements of architecture or landscape on the banks of the lake, such as the temple of lake goddess—Ganggavva, or the tree planted on the occasion of the visit of the Norwegian Prime Minister. These repositories of meanings become very important in establishing an identity for a space that manifests as a particular place and creates the idea of relatedness between the human being and her surroundings. The same is true of those elements of nature as *prakṛti*. While nature is whole, we cannot relate to the whole, we can only relate to its various elements—trees, rivers, landscapes, animals, and so on. Relationships are not to be seen as natural dependency and biological interdependencies between these various elements of nature. Instead, understanding relationships as the relatedness of the human being to nature through the process of "making sense of" would give an alternative perspective. The "making sense of" the world within the view of Indian thought is interesting and becomes a rich source of meanings that help us understand nature.

Merchant (2004, p. 223) proposes a new environmental ethic—a partnership ethic. She suggests that it is based on the ideas of a "viable relationship between a human community and a non-human community in a particular place, a place in which connections to the larger world are recognised through economic and ecological exchanges." In Indian thought, we already have a similar form of this mutual relationship and obligation within the concepts of *karma* and *dharma*. How might we articulate these principles to support an ecologically-relevant ethics? As she suggests, a mutually beneficial situation requires that both people and nature are acknowledged as actors (p. 223). We have seen that nature cannot be a moral agent in the Western traditions. Merchant proposes that the concerns of nature be brought to the table on discussions related to any project or intervention in nature. Nature should be accorded a voice in all our meetings. As an equal partner with human beings, consensus and dialogue should be attempted at all times keeping the interests of both humans and nature. She writes:

> A new ethic entails a new consciousness and a new discourse about nature. Living with and communicating with nature opens up the possibility of non-dominating, non-hierarchical modes of interaction between humans and nature (p. 229).

She adds that mechanistic conceptualisation of nature is replaced by the position that nature becomes a subject. The voices of human being and nature would both find expression in such an ethic (Merchant 2004, p. 229). To accord voice to nature requires is to humanise nature, to give it equal moral standing. The current paradigms of ethics cannot account for moral standing except through invoking concepts such as intrinsic value. Categories we have seen in Indian thought such as *prakṛti* and *dharma* point us to ways in which nature can be brought to the table.

As we have seen in earlier chapters, nature in Indian philosophy can have *dharma* as its voice. The humanising of nature is a common theme through many narratives. Even when the earth is a divinity, she speaks with the voice of nature in a dialogue with King Pṛtha. In another episode from the *Māhābhārata*, the deer appear in a dream to Yudhiṣtira asking him not to deplete their numbers in the forest. Trees, parrots, elephants, mountains, and rivers all have voices in Indian thought. They speak for their own *dharma*. While on one hand, it is easy to dismiss these voices as imaginations of pre-modern peoples; on the other hand, understanding this as the "voice of *dharma*" provides us with an idea of what we think nature wants. In a verse regarding the stubbornness of his heroine, Kālidāsa, the great poet remarks that the Gaṇgā water would not flow upwards. The river wants to flow downwards that is its *dharma*. For example, on the discussion during the meeting for a large hydroelectric project, if we ask what the river wants, it would articulate its *dharma*, which it that it wants to continue to flow downwards. If we listen to the voice of the river—the riparian rights of people and animals downstream, seasonal variations, floods, and the erosive action of a river whose work is also to replenish the silt in the plains—all of this would have to be taken into account, not just the socio-economic cost-benefits of the dam.

The principle of ecological ethics in Indian thought is fundamentally based on the unique, internally relational, substantive, yet functionally differentiated constituents of the universe. These elements find themselves expressed in alternative discourses of meaning making of the people, whose interaction with the everyday world is often given by narratives rather than by any understanding of "facts" or "concepts". This world view is combined with a strong normative principle of action, where being and function are interrelated. To be human is to be within the realm of both *ṛta* and *karman*, and this means to be related to every other created existent in the world.

It is within this cosmic process of relatedness between created beings and the environment that the main eco-ethical themes discussed in this book can be contextualised. The relationship between the substances and their manifestation constituted by *guṇa*s provides for a framework of evaluation that does away with a categorical view of conservation, and replaces it with a relational view. We as human beings conserve not because we are different and separate from "nature" but because we are also *prakṛti* and relatedness inheres in everything as *dharma*. The same *dharma* inherent in human beings as members of a created cosmos (*nisarga*) is embodied in different bodies and en-worlded by *loka*s expresses itself as care, trusteeship or being non-violent and in being a shelter towards the earth, the other beings, and the environment. In this case, it does away with the focus on debates on sentient–insentient distinctions of environmental ethics and shifts our attention to the agency (*kartṛtva*) from the arguments about considerability of moral action. Again, as human beings, we are embodied in the functions of being related to the processes of natural resource degradation, by being an efficient cause. Conservation as slowing this change requires us to fulfil the *dharma* of being cause in the matter of slowing down the processing and consumption of raw materials. Finally, as embodied beings connected to the objects of experience through the relationship of disinterested action, we must act morally towards ourselves and with restraint towards the sense objects that envelop us as nature.

The recent debates surrounding conservation in India have raised issues regarding the kind of knowledge that should form the basis of management of our natural heritage. Many conservationists are clear that biodiversity cannot be conserved in isolated pockets of "wildernesses". The depletion of natural resources and the new challenges of population and globalisation have only added to the ecological crisis. Leading environmentalists have suggested recourse to mitigation and adaptation as two practical methods of handling the global crisis.

Before there is a real pragmatic change in action or policy, there needs to be a conceptual transformation of the way human beings perceive nature. It has therefore become imperative for philosophers to examine the different aspects of the relationship between human beings and nature. As a discipline, environmental philosophy is still in its early stages of development compared to other branches of philosophy. Colyvan (2007) states that there are many interesting philosophical issues associated with the science and policy of conserving our natural environment that go beyond the scope of just environmental ethics. I have therefore focused on the broader questions and ideas around the conceptualisation of nature in Indian thought.

Such shifts have to occur through the engagement with various forms of narratives and texts found with or culture. For instance, as pointed out in Chap. 9, ecocriticism, a discipline that has its origins particularly in American literature, is yet to develop concepts that can address non-Western, non-English literature. When we try and understand nature within pre-modern Sanskrit literature in India, where the influence of Romantic Movement is completely absent, we find that to refer to these compositions as nature-centric or human-centric is also difficult. A more nuanced way of understanding nature in pre-modern poetry could be through the two categories of action and natural behaviour of the non-human world. As we have seen earlier in poetry of Kālidasa, the cloud's activities in the rainy season are described rather accurately.

The discourse of *dharma* and *karma* of nature in literature is one of the ways in which we could analyse nature that is inclusive of the human being. In literature, the non-human is included in the human world. Nature is seldom passive in these narratives; nature participates in the human world by being itself, active as nature. Winds blow, clouds rain, the forest fires burn, and mountains stand tall. All these are *dharma*, the appointed duties of natural things or the order of nature. *Karma*, as actions of nature objects, is completely instinctive and in sync with their *dharma*. Yet again, it is through these categories that the feminisation of nature or the silencing of nature takes place.

The descriptions of human engagement with nature are also based on karma and dharma. The adaptation to seasons and landscapes and the ethical and moral attitude to the non-human part of the world are all based on activities and the order of nature. The literature discussed in Chap. 9 in many ways reflects these important conceptualisations from Indian philosophy.

To summarise, I posit that there are three main frameworks in Indian conceptualisations of nature for proposing an ethical relationship with other "created" beings around us: firstly, the internal relatedness created by the concepts of *prakṛti* (as constituted by the *guṇas*) as well as *nisarga* (as levels of created beings and functions) that places human beings "within" nature, yet allows for eco-moral responsibility in the context of *dharma* and moral action. The second framework is that of an embodied, en-worlded relatedness to the planet we live on, understanding nature as being entrusted to us, in which trusteeship and ethics of non-consumption through disinterested action play an important role. Finally, a phenomenal relatedness framework is created by the concept of human beings as caregivers to nature, from another perspective of human non-injury to the various beings (including the environment beings) as "beings that sheltered" by us, their shelterers. This relatedness is to be interpreted as more of an existential, an "allowing to be" relationship. This has important implications for environmental and conservation ethics. Conservation therefore has to be relational and not oriented towards a particular being or object or species or a particular space. Within this relatedness of dependency and welfare, which are not steady states of affairs, it is clear that at various point of time, various beings can support or be supported by other beings.

References

Berkes, F., Kislalioglu, M., Folke, C., & Gadgil, M. (1998). Exploring the basic ecological unit: Ecosystem-like concepts in traditional societies. *Ecosystems, 1*, 409–415.

Bijoy, C. R. (2006). Kerala's plachimada struggle: A narrative on water and governance rights, *Economic and Political Weekly*, October 14, 4332–4336.

Casey, E. S. (2001). Between geography and philosophy: What does it mean to be in the place-world? *Annals of American Geographers, 91*(4), 683–693.

Colyvan, M. (2007). Environmental philosophy: Beyond environmental ethics. Extract from a lecture delivered to the Arts Association, Sydney on October 2007. [Online] http://homepage.mac.com/mcolyvan/papers/envphil.pdf. Accessed on 29 March 2010.

Deustch, E. (1989). A metaphysical grounding for natural reverence: East–west. In J. B. Callicott & R. T. Ames (Eds.), *Nature in Asian traditions of thought* (pp. 259–267). Albany: State University of New York.

Guha, R. (1989). Radical American environmentalism and wilderness preservation: A third world critique. In J. B. Callicott & M. P. Nelson (Eds.), *The great new wilderness debate, a collection of writings defining wilderness from John Muir to Gary Snyder*. Athens: University of Georgia Press.

Larson, G. J. (1987). Conceptual resources' in South Asia for 'environmental ethics' or the fly is still alive and well in the bottle. *Philosophy East and West, 37*(2), 150–159.

Merchant, C. (2004). *Reinventing Eden: The fate of nature in western culture*. NewYork: Routledge.

O'Niel, J. (2003). Meta-ethics. In D. Jamieson (Ed.), *A companion to environmental philosophy* (pp. 163–177). Oxford: Blackwell.

Trudgill, S. (2001). Psychobiogeography: Meanings of nature and motivations for a democratized conservation ethic. *Journal of Biogeography, 28*(6), 677–698.

Index

M. Baindur, *Nature in Indian Philosophy and Cultural Traditions*,
Sophia Studies in Cross-cultural Philosophy of Traditions and Cultures 12,
DOI 10.1007/978-81-322-2358-0